DATE DUE

NOV 07 2011	
Dec. 10, 2013	

Diamond Stories

Anthropology of Contemporary Issues

A Series Edited by Roger Sanjek

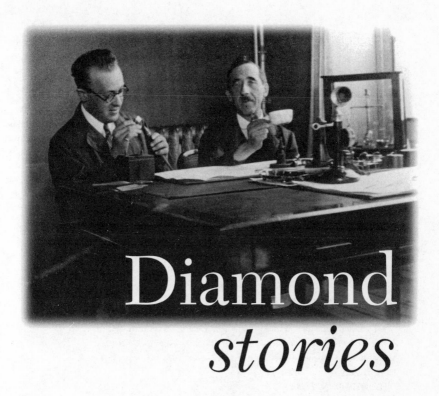

Diamond
stories

Enduring Change on 47th Street

RENÉE ROSE SHIELD

CORNELL UNIVERSITY PRESS

Ithaca & London

First published 2002 by Cornell University Press

Printed in the United States of America

Library of Congress Cataloging-in-Publication Data
  Shield, Renée Rose.
     Diamond stories : enduring change on 47th Street / Renée Rose Shield.
        p. cm.—(Anthropology of contemporary issues)
     Includes bibliographical references (p.).
     ISBN 0-8014-3989-2
        1. Diamond industry and trade—New York (State)—New York—History.
     2. Jews in the diamond industry—New York (State)—New York—History.
     3. Jewish businesspeople—New York (State)—New York—History. I.
  Title. II. Series.
     HD9677.U53 N77 2002
     381'.4282'097471—dc21
                                                    2001007138

Cornell University Press strives to use environmentally responsible suppliers and materials to the fullest extent possible in the publishing of its books. Such materials include vegetable-based, low-VOC inks and acid-free papers that are recycled, totally chlorine-free, or partly composed of non-wood fibers. For further information, visit our website at www.cornellpress.cornell.edu.

Cloth printing   10 9 8 7 6 5 4 3 2 1

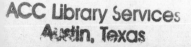

In loving memory of Shmiel and Moishe—
with them an era passes.

And to the glory that is New York.

Contents

Preface

The story of Jews working in the diamond industry in New York tells how they mix in and stay apart, how they are open in important ways to mainstream American "culture," and how they also adapt to new times in ways that are both modern and traditional, indeed ancient. Though this process is common to all ethnic groups, the role of diamonds in the history of Jews adds a special twist. Here is a story of transformation and re-creation through change. The diamond, a pebbly object transformed into a twinkling, astronomically priced jewel, has allowed Jews to transform themselves from rejected refugees of one country to respected businessmen of another. Time after time, they have occupied an important but vulnerable niche in a process that has repeated itself over and over throughout the centuries.

Some Jews, inheritors of this long tradition, center in the city of New York, where they integrate themselves fully in the twenty-first century while still forming a sequestered world within a few blocks of midtown Manhattan. Thoroughly modern, their identities are linked to biblical roots.

Today this global industry nimbly blends two worlds: one of informal handshakes, a Talmudic system of resolving disputes, and the friendly banter of *schmooze* [idle talk], another of high technology, including laser cutting, computer imaging, and Internet shopping.

This book looks at who these people are through the lens of a person linked to and separated from this world. As a cultural anthropologist and the niece and cousin of several New York diamond manufacturers and dealers, I asked them to allow me entrée and they graciously agreed.

♦

I grew up an assimilated and secular Jew in Connecticut's Fairfield County, only an hour from my transplanted European relatives in New York. How different our families were—yet, those differences melded amiably with the links between our families and delighted me as a child. Their homes held some of the same artifacts that mine did—vestiges of the European home they had all left—but in their New York homes, Yiddish was spoken, and Jewish rituals were more elaborately, more *properly*, carried out. My cousins knew all the Hebrew verses in the Passover songs and could chant the prayers in the Seder dinner while we ventured to chime in when a passage was faintly recognizable. When we went to visit, one crystal dish brimmed with small, smooth discs of bitter chocolates and another held the delicately flavored Dutch coffee candies, Hopjes. One aunt always had elegant Dove soap on the upstairs bathroom sink, and I somehow thought its glorious smell was European. My cousins drank coffee, and we guzzled milk. They had the same Jewish books we did, but they had more of them, and many had Hebrew letters embellishing their spines.

The similarities and contrasts between our families parallel the continuities and changes that coexist in the diamond business. As our families blended the old and the new, each in its own idiosyncratic mix, the twenty-first century diamond business retains ancient traits while it confronts and courts globalization. Each synthesis is complex and evolving.

♦

Human beings, simultaneously calmed by assurances of continuity and excited by and fearful of flux, struggle to understand how to mediate oppositions, grow but not grow old, stay the same yet evolve. Anthropologists know the positive glow that people construct about a glorified past, and diamond traders sometimes indulge the idea of a mist-shrouded past, too. As the diamond traders of New York adhere to traditional ways of manufacturing and trading their goods, they also mold aspects of their work and their lives in accordance with new developments—political, technical, economic, and cultural.

And there is individual aging. Many of the diamond traders and manufacturers in this book, including the two uncles who inspired it, became adults in the diamond business and became old in it. I encountered a widespread wish by many diamond traders to continue to work into extreme old age: they hope that good health and good business will allow them to keep working. Many men in their 80s, some in their 90s, keep plugging away in this most social of businesses, both buoyed and challenged by it.

♦

Jews and precious jewels have been a match since the first millennium, when early trade routes spread from Southeast Asia to India and the Mediterranean, linking Jewish centers from the Ottoman Empire to the Netherlands. Jews were separate from the people among whom they lived and were often considered strange. Subject to numerous prohibitions, Jews lived, at times, precarious and stressful lives. The jewelry trade was open to Jews who were restricted from many occupations and professions, and early trade and guild restrictions did not yet apply to them. Because banking and commerce were fields in which they were allowed to work, they lent money, were pawn dealers, and appraised and repaired jewelry. Diamonds and pearls gave Jews access to royal courts throughout Europe, Egypt, and the Mediterranean. Jews traded, they were persecuted and kicked out, and they moved on. Their mobility made them multilingual and cosmopolitan. Like their books—including the Torah—that could be carried from place to place and did not need to be dependent on only one place, diamonds were portable and suited to the diasporic life of the Jews. When they had to flee, their tiny diamonds could be ingeniously hidden—stitched into corsets and shoulder pads of clothing, tucked into the bottom of toothpaste tubes—diamonds were often critical to survival.

◆

Diamonds are perceived as extraordinary, and 47th Street stands out within New York as a place apart, an exotic enclave. The hundreds of bearded and black-robed Hasidim loping across and down the street, rushing to do business, encountering acquaintances, speaking rapid-fire Yiddish, and bounding into another narrow doorway of another diamond exchange creates a look of complete separation from sleek 5th Avenue at the corner. But that look belies the intimate and twined relationship between 47th Street and 5th Avenue.

Though the district appears dominated by the black-hatted Hasidim, it is complexly mixed. Ethnicity has changed gradually but markedly in the district over the years, despite a dramatic increase in Hasidic Jews. Though most of the traders are Jewish, their countries of origin are diverse, and varied expectations and styles add a lively component to individual personality differences. Furthermore, Indians, Chinese, Thais, Filipinos, Japanese, Italians, Israelis, Iranians, and others—only some of whom are Jewish—trade actively (and sometimes in Yiddish) and mix it up.

In this way the district is a typical New York neighborhood, apart from and conjoined with New York, where one block can be unmistakably Dominican, another completely Russian; where neighborhoods are both distinct and blurred; where they seem to stand apart, flaunting separation and yet composing the tightly knit fabric of the city (see Sanjek 1998). The

United States is not Americanizing the world. Global cultural flows are at work, not limited by geographical underpinnings, and they influence how people link and divide themselves in relation to one another individually and in groups. While global, these processes are very much locally determined. In this respect this ethnography is a story about New York perhaps as much as anything else.

◆

A book such as this relies only so much on archival and reference materials. The heart of the book comes from people who were willing to share their diverse, sometimes contradictory, and often controversial perspectives with me. The diamond district is in many ways a close-knit community, stereotyped by outsiders as insular and secretive. Though I could not promise that the individuals who appear in these pages would not be identifiable to one another, I assured them that I would use pseudonyms. In some cases I changed certain of their characteristics to make them less recognizable. I make an important exception with my uncles: I use their real first names to introduce them and to describe how I did the field research.

Because of my attempt to respect the privacy of those who consented to talk with me, I cannot name most of the people to whom I owe heart-felt thanks. These are, first and foremost, the numerous diamond traders in New York with whom I held long and informative conversations. Their good-natured generosity was remarkable. They offered their knowledge, their time, and their opinions. Their humor and warmth made working among them a pleasure.

I shudder at my *chutzpah* in infringing on my uncles and cousins and their families, but they just shrugged and took me on, month after month after month. They explained, they were open, they filled me in, they introduced me to others. They let me accompany them, and they let me be so I could watch and ask questions. They put me up, and they put up with me.

I am indebted to other diamond traders in my family, who though now deceased, have informed and enlivened these pages, especially my grandfather Chaim, who started the family business, and his oldest son, Aron.

I am grateful to officers of the Diamond Dealers Club of New York who gave me access to the club premises to do this research. Personnel from the DDC, the American Diamond Information Association, the Gemological Institute of America, the trade magazine *New York Diamonds*, Larry Feldman, and the staffs at the libraries at the YIVO Institute for Jewish Research, Brown University, and my town of Seekonk, Massachusetts, were invaluable in providing information and securing reference and archival materials. I am indebted to Charlie Hollander for letting me use his mother's diary, and my son, David, for translating it from its original Flemish.

Early and continued encouragement came from my parents, Anne and Gilbert Rose; my in-laws, Betty and Rocco Bruges; my brothers and sister, Dan, Cecily, and Ron, and their families; and my other relatives.

I offer sincere thanks to families, friends, and colleagues for reading and feedback: Stanley M. Aronson, William O. Beeman, Carol Cummins, Barbara and Walter Feldman, Ellen Fishman, Betty J. Harris, Rachel Kaufman, Gisele Kaufman, Abe Kaufman, Arthur Kaufman, David Kaufman, Jacob Kaufman, Isi Kaufman, Abraham Kaufman, James Kaufman, Susan Kertzer, Judy Lewis, Anne McDonald, and Gillian Rogell.

The late George Hicks steered me early on to Abner Cohen and was almost as excited as I was by this subject. I am grateful for the helpful comments of an anonymous reader and for the expert guidance of Cornell University Press, notably Roger Sanjek, Fran Benson, Peter Agree, Candace Akins, Priscilla Hurdle, and copy editor Karen Bosc.

Sonja, Aaron, David, and Lily were the greatest kids a writing mom (or any mom) could ever have throughout this long process. Lily—so worth waiting for!—literally grew up with this book. My husband, Paul, lovingly and unflinchingly believed that this endeavor was worth doing—and by me. I know how lucky I am.

My uncles, Shmiel and Moishe, quickly overcame their initial reluctance to embrace this project warmly and strongly. Moishe's approval of my first draft reassured me that I was on the right track. I miss them greatly and wish they—and my mother—were able to read the finished product.

Finally, while it is customary to note that loved ones showed forbearance and patience during the time the author labored over her manuscript, it is *because* of my family that I have taken so many years to produce this book, flawed though it may be. I have woven my career and writing around my family, and so I insist they share the blame and the credit for it. That they have been the central project of my life has been the best choice I ever made.

In this book I have included some Yiddish and Hebrew words. As much as possible I have attempted to adhere to the standard YIVO transliteration system, while also reflecting the actual usage of these terms. In some cases I have relied on nonstandard variants of the words, such as in proper nouns and in material quoted from other sources.

Diamond Stories

[1]

Introduction

Adding glitter to the cynical and weary world, and revealing no hint of their earthy origins, diamonds are everywhere. This supreme symbol of heterosexual love glimmers on slim fingers and graces delicate as well as ample necklines.

Vast changes in the traditional diamond industry vie with incessant pressures from the modern world. Once scarce and sacred in the earliest days of diamond discovery several millennia ago, diamond abundance today is concealed by the sophisticated advertising message of precious exclusivity. Increasingly in recent years, the tension between constant evolution and persistent continuity has created an uneasy shifting balance. Today, as in the past, venerable traditions mingle almost seamlessly with new technologies. Today, as in the past, handshakes, Yiddish, and trust still close the multi-million dollar deals.

◆

The marketplace itself retains important traits from the past. Haggling face-to-face in the club high above 47th Street, men's voices are vociferous and volatile. Fingers waving, heads shaking, and calculators punching out discounts and commissions punctuate the fractious atmosphere. Sometimes, emerging from the din, hands slowly extend to clasp and signal agreement on price and terms.

How many observers of the 47th Street diamond scene have, like me, sensed its affinity to a Middle Eastern bazaar? Fascinated, I watched the following exchange early in my field research, not comprehending.

The broker, a big stocky man, gives a ring back to its owner, Chaim.[1] He says he almost had it sold, but too bad! The wife didn't want it at the last minute. She wanted a Bentley instead, ha ha ha, HA! These women. Also, . . . maybe the GIA certificate isn't really right. Maybe, he suggests, nudging Chaim, he should contest the certificate with the GIA. Chaim looks neutral as he accepts the ring back. No comment. No hint in Chaim's blank face as he unfolds the parcel, takes out his loupe, checks that it's the right stone, closes the parcel, inserts it back into his folder, and then back in his pocket, face closed. A slight nod to the man, message received, he can go now.

After the broker left, I asked Chaim: What was that? What happened just then? Oh, Chaim said simply, turning to me, his shoulders even, his demeanor cool. It's obvious: everything the broker said was bullshit. The man couldn't sell the ring, that's all. It doesn't matter why. The broker felt he had to make up the stuff about the Bentley rather than simply say the deal fell through. There is so much bluster and charade. It'd be nice if people could say it straight and not go through excuses and storytelling. That talk about the certificate—same story. An empty attempt to compliment Chaim. He doesn't need the broker's stroking. He wants the sale! Such chatter is a waste of time.

The Diamond Dealers Club (DDC), the central marketplace for wholesale diamond dealing in New York, was the scene for this and for countless other unchronicled transactions that proceed every day. The DDC has, according to one writer, the "aura of a nineteenth-century Middle Eastern bazaar under the light of TV screens and the whir of fax machines" (Kanfer 1993b:viii). The parallels as well as the differences between the bazaar and the diamond trade are intriguing and stark.

Anthropologist Clifford Geertz studied the Moroccan bazaar in the town of Sefrou, a hub in the ancient caravan economy that linked a series of markets in Morocco (1979). The male-dominated, Islam-permeated bazaar encompassed different ethnicities, including Jews. "Moroccan to the core and Jewish to the same core . . . heritors of a tradition double and indivisible and in no way marginal," the Jews here were both integrated and separate, demonstrating the paradox of a "curious, just-the-same, entirely different" relationship with the Muslim community (1979:164–5). Like every other subgroup, they differentiated themselves according to their trade and product specialties, including gold working, tinsmithing, kosher butchery, shoe-

---

[1] I use excerpts from my field notes liberally throughout the text. The names are pseudonymous. Most of these notes are descriptions of incidents and close paraphrases of conversations in which I participated or that I observed. My use of quotation marks means that I am certain of the exact wording. Most of the quoted material derives from transcripts of tape-recorded interviews.

making, or peddling. As in other places these Jews were separate and made themselves essential at the same time. They operated their own religious and secular organizations and adjudicated their commercial disputes.

Picture the cacophonous setting of the *suq*, where the main task of all traders was to discern shrewd prices and values of the market goods they were exchanging. "The real pushing and shoving is done with talk," Geertz commented, describing the disorder as "hundreds of men, this one in rags, that one in silken robe, the next in some outlandish mountain costume, jammed into alleyways, squatting in cubicles, milling in plazas, shouting in each others' faces, whispering in each others' ears, smothering each other in cascades of gestures, grimaces, glares—the whole enveloped in a smell of donkeys, a clatter of carts, and an accumulation of material objects God himself could not inventory . . . sensory confusion brought to a majestic pitch" (1979:197).

Lacking brand names, trademarks, advertising, and systematic information about products, and confronted with huge price disparities, the trader had to decide whom and what to believe. "Sorting lies and nonsense from the reasonable and the real is the information problem," Geertz reported, and haggling was intense. Business disputes were settled by whatever local witnesses were reliable.

Beyond his goods, a trader had his reputation. He traded with only a select few of the people in the market, and they could judge his information and honor through the history of his transactions with them and others over the years. Among men simultaneously joined and competing, bargaining by rules of etiquette including tea-drinking and effusive rhetoric, everything was negotiable, especially credit terms and repayment schedules. Keeping the players playing with one another was the most important principle.

The similarities between the diamond club and the Sefrou *suq* stand out: few women, religion interwoven in the customs and rubric of the trades, extraneous "noise" clouding the search for reliable information, haggling bound by custom and rhetoric, negotiation about credit, the use of brokers, oral contracts, mediated disputes, the supreme importance of reputation. As in the bazaar, diamond traders minimize their risks by limiting their trading partners to a small number of well-known and reliable associates to whom they show their precious goods over and over again.

In both marketplaces contracts are oral and are considered binding and legal. Disputes are mediated by trusted authorities in each group, and virtually no recourse is made to outside governmental or other authority. A damaged reputation can ruin a trader. In addition, *suq* and diamond traders usually do business with one person at a time. Sefrou's go-betweens helped connect Sefrou hierarchically with other trading towns. Diamond brokers connect traders within New York as well as between cities as they travel with goods to trade shows and auctions and create links throughout the world.

Information—noisy, distracting, constant—creates a busy and winding maze in the DDC as in the *suq*. Though the diamond market is aided by increasing standardization that has affected the industry profoundly, a trader still spends time probing the little nuggets with a handy loupe (the jeweler's magnifying glass that fits on a thumb and often hangs by a cord or a chain around the neck) to detect fluorescence, imperfections deep within the stone, deficiencies in the cut, and artificial embellishments, as well as to confirm authenticity. Though large diamonds increasingly come with grading reports from the Gemological Institute of America (GIA), the traders still rely on their own judgment about the value of the diamonds, regardless of whether a stone has been graded. While De Beers in 2001 launched plans to brand packets of stones with identification information—largely to screen out the illegal, "dirty" diamonds that are financing African civil wars—in general there is no Chiquita, Gap, or ILGWU label on a diamond to confer public authenticity, quality, or origin.

So it is not insignificant that the diamond market, like the bazaar, is rife with the buzz of peripheral, confusing, extraneous chatter, of obfuscation, prattle, banter, jokes, cajoling, and *schmooze,* all of which conceals or obscures the information the trader needs. New diamond traders must become adept at distilling that essence from the clatter and tumult of great amounts of entertaining, diverting, and insistent talk. This skill of discernment is not necessarily intuitive; though some are "naturals," such skill is for the most part learned by experience, careful consideration, and honed judgment. It is assisted by the seasoned practice of trusted others who know and can be relied on to share their counsel.

How do traders sort the noise and evaluate bluffs? A dealer told me that sometimes a person considering one of his stones holds his reply a few weeks and then says ominously, "There's competition." He laughed as he continued: "They try to scare us by saying there's competition, somebody else has another stone that's cheaper [that they'll buy instead]. So I always figure when they say there's competition, that means the lady likes *our* stone, and they're just trying to scare us. So I say, forget about it! Bring our stone back. Sell to the competition." But the dealer also knows he will give a small reduction in price to keep the business going and to maintain the good will. Good advice to a would-be diamond trader, these lessons of sticking to the bottom line, of suffering the talk to get to the essence, and remembering the need for continued relationship that underlies the entire enterprise.

Similar machinations amid the overblown talk occurred in the *suq,* too, but the differences between the two markets are profound. Diamond traders abide by many traditional and unwritten rules and by trade, New York state, U.S., and international bylaws, laws, and regulations. The Sefrou bazaar appeared to have no overarching regulations beyond binding customs. Also, the one basic commodity in the diamond market contrasts with the multitude of products in the *suq*. Though membership in the diamond-trading club is exclusive and bound by extensive written rules and bylaws, barriers to trading in the *suq* were few in contrast. And while individual expertise for judging the merchandise is necessary in both markets, the increase in standardization in the diamond market—unheard of fifty years ago—has changed the tenor of the marketplace fundamentally, to the extent that independent trader expertise, experience, and judgment are less necessary than previously.

There is no advertising in the *suq* and much less standardization in weights, brand names, price quotes, and consumer guides; prices vary greatly as a result. By contrast, fifty years of "A Diamond Is Forever" campaigns made luscious diamond advertising prominent throughout the world. The bazaar stands at the far end of the local market economy, whereas the diamond market is rooted at the other extreme, with international cartel control of this commodity responsible for the general stability of diamond prices worldwide.

Are such markets with this kind of informational maze truly unique? Comparing simpler societies with our own, neat divisions between "us" and "them" are usually hard, if not impossible, to draw. Elements of bazaar economies may be present in all kinds of marketplaces. Many systems of exchange may involve "bazaar-style information searches" with unstandardized prices and quality of goods, as Appadurai has suggested (1986:43). Though vast differences separate the two marketplaces, certain bazaar features strenuously connect them.

Though the Moroccan bazaar linked far-flung towns together with trade, the globalism of the diamond industry surpasses most business in the world and stretches back in time. Since the early days when diamonds were considered valuable, magical, and replete with symbolic significance in places like India, East Asia, Russia, and Europe, diamonds began to be sought along far-reaching trading routes, and these routes helped forge some of the first linkages between these societies. In the last century the trade became more fervent, concentrated, and controlled as diamonds were discovered and mined in increasing numbers of places. Modern communications in the last few decades—and most stunningly, in the last few years—has increased the international agility of present-day diamond trading, enhancing the coordination of the diamond bourses worldwide and underscoring how vast and complex the diamond network has become. Trading in this insular yet global business dramatizes how people can integrate themselves within a society but remain apart at the same time in a world growing more complex and intertwined every year.

◆

This book begins with an introduction to the diamond itself. Chapter 2 provides a background context that anchors the ethnography by first describing what the diamond is physically. Next, I relate the complex geopolitical history of the diamond. The several strands of this story are intertwined and include the history of mining diamonds, the origins and rise of the De Beers diamond cartel, the role of Jews in the diamond trade, and the role of diamonds in conflicts, particularly in recent decades. I add reaction and commentary by current diamond traders in New York to various historical sources and to newspaper and financial accounts that I include here. This contextual material is used as backdrop to launch the ethnography so as not to detract and distract from the central players of this work, the present-day New York traders.

Chapter 3 relates how the ethnographic field research was conducted by describing the setting, my methods, and my perspective. The chapter is intended to orient the reader in the district and situates the author in my various roles—as anthropologist, niece and cousin to fellow diamond manufacturers and merchants, and general outsider—in relation to the traders.

Chapter 4 provides a perspective on New York's central role in the worldwide diamond trade as well as the growing presence of the Jewish Hasidim among the ranks of the traders.

Chapter 5 describes the mechanics of trading in the New York marketplace of the Diamond Dealers Club and introduces the kinds of changes that are penetrating the district.

Chapter 6 explores the widespread use of the GIA grading reports and the price sheets promoted by Martin Rapaport—developments that are considered to be seismic changes in the industry and that have led to increased standardization and competition.

Chapter 7 concerns other important social changes. The term "family business" aptly describes the perhaps hundreds of small enterprises peopled by reliable and trustworthy family members both local and distant. Will children continue to enter the family businesses when a plethora of other choice beckons and the trade seems precarious? Whether the family business will break down under the rapid consolidation of modern globalization or be a model for successful entrepreneurship remains to be seen.

The position of women in the trade has also undergone change. From early days, women in the diamond trade have not only heard the stories, discussed the business, offered advice, partnered with their husbands, and groomed their sons, they have also worked in the wholesale and retail trade as sorters, secretaries, polishers, wheel turners, saleswomen, office workers, brokers, dealers, jewelers, designers, manufacturers, and buyers. Because men have dominated the writing and research of Jewish texts traditionally, the activity and influence of women have been little noticed or documented.

Female researchers, armed with anthropological theory and using oral histories and personal experiences, have begun to examine how Jewish women work for pay, are partners, and contribute in quiet ways—and how they also resist traditions and others, sometimes as they often become more powerful with age.[2] Religion, particularly Jewish Orthodoxy, has been a large factor in keeping women more or less out of actual, if not symbolic, sight within the diamond trade. How do these women compete and cooperate with the bearded Hasidim, ultra-Orthodox in their Judaism, forbidden from face-to-face contact with women not their wives or sisters?

Like the Chagga people of Kilimanjaro in Tanzania who lived their entire lives in proximity with one another, embedded in a dense web of social and financial interdependencies—a phenomenon termed a "life-term social arena" by Sally Falk Moore (1978)—many diamond traders in New York exist in a network whose kin and community solidarity endures over time. Despite ethnic diversity and mobility, the continuity of personal contact cements, challenges, and modifies relationships constantly. Traders vie with and sustain one another, often into advanced old age—if they are able. Though the work is fraught with risk and unpredictability, it provides social supports that cushion some of the difficulties of aging. Many desire to stay in the business until death. Work defines them, gives meaning to their lives, structures their days in positive ways. Their words tell how they derive meaning from continuing in the trade. With increased economic pressures will people retire earlier or will they continue to work in their older years? Though most people over age 75 in the United States are retired, diamond traders who are healthy provide vivid contrast.

In Chapter 8 I turn to the process of arbitration in the trade to examine how this way of mediating disputes among diamond club members is uniquely suited to the industry. In many ways arbitration represents the crowning achievement of the diamond business and is emblematic of how trust and reputation are the stellar symbols of the trade. I examine how this "courtly system of arbitration," as one diamond dealer reverently called it, endures even as the American legal system has made inroads into it.

The system is based on Talmudic principles of common sense and fairness that are inculcated from an early age, especially among the Orthodox

[2] For sources on women in the diamond business and Jewish women in general, some sources are Patwa 1989, Davidman and Tenenbaum 1994, K. Sacks 1979, M. Sacks 1995, Sered 1994, Rosen 1992, Belcove-Shalin 1988a, 1988b, 1995, Weinberg 1988, El-Or 1994, Hyman 1994.

and the ultra-Orthodox. The writings are thus well-known and are strenuously argued in groups. The prescriptions to keep one's word, to pay debts and not buy stolen goods, to not speak ill of others, and to resolve conflicts by arbitration are enshrined in the Talmud. Severe judgment is reserved for those who impugn another's good name.

The arbitration system continues a process begun centuries ago in the *kehillah* [community] in Jewish society. Because Jews usually lived in lands in which they were persecuted and often isolated, their distrust of outside authority became an ingrained cultural response. The *kehillah* consisted of a leadership group of important men and a religious court, the *Beth Din*. In general, the reigning powers recognized the authority of these courts and imposed little interference. These organizations continued in the United States for much of the twentieth century, when adjustment to an American way of life was difficult. The use of familiar religious law and Yiddish helped communicate a trustworthiness that many did not find in the secular state, and it kept internal Jewish disputes from the shame of public display. Similarly, arbitration in the diamond trade sends a symbolic message to the New York courts to not transgress, and it allows the Jewish traders to reflect back onto themselves the defining message "This is ours; this is us; let's settle this among ourselves."

Chapter 9 concludes the book by considering how changes in the industry—especially globalization—have affected the trade. In this chapter I offer ideas that link the changes in the diamond industry to those occurring worldwide in other businesses.

◆

Throughout these pages I weave diamond stories that I was told. The stories and the people in their own voices offer glimpses into various aspects of their work, experienced with unique perspectives. Some of the stories are very old, and some are generated *de novo* on the spot. The stories evoke the ancient and universal performance of talk in the *suq*, where individuals show their wares, buy and sell, reminisce, test one another, compete, and connect. The stories define but do not limit the people of these pages. Their words construct how they see their world and how they act in it. Their words celebrate who they are, and are the medium by which they maintain their differences, defy competitors, and slyly survive. Through their words and through their craft, they make their lives while they make their living. As individuals in the present creating community through one another, they link to the past, face the future, and together endure change.

◆ ◆ ◆

[2]

Diamonds, Jews, and the World

Trucks cram New York's 47th Street, and people dart between them to cross to the other side. Booths of glittering wares sparkle behind storefronts that advertise great deals. People walk quickly. Men wearing cardboard signs press their flyers into the arms of passing pedestrians. Small glass storefronts brimming with jewels alternate with dingier entryways along the block. Some of these open into mall-like areas of tiny booths that swarm with people. Other doorways lead into narrow hallways with elevators that ascend to scores of offices above. The storefronts on the street level, like the interior booths behind them, display dozens of jewels, necklaces, bracelets, and rings, new and antique. Signs at the front of the stores complement the multitude of signs hanging from the entrances. Tourists ogle and marvel, heady with the abundant glitter; they have heard of this place. Quiet couples holding hands gaze in front of one display, then another, contemplating whether to venture inside. Middle-aged women hurry by, glance quickly, go in; they know this place. Police, uniformed and not, patrol unobtrusively.

Many men clothed in the traditional dress of the Hasidim walk down the street, march quickly in and out of doorways leading to and from offices on the street level and above, cell phones pressed to their heads. Various men, some in long, white beards, greet each other as they cross the street or pass along the sidewalks. Some stop and talk quickly. They've left one another messages in their offices; they count on seeing one another here or upstairs. They remind one another of payments due, make inquiries about goods they're interested in, question how the goods they've consigned are faring, renew promises to return goods, and provide news about prospective customers. "Bring me the four eighty-six," one says to the other. "I have some-one, may be interested." "Do you have a one and a half pear-shape, VVS, j-k color? Bring it up." They arrange to meet later, either in the club upstairs or in one of their respective offices.

The trucks make deliveries; they block the street as cars press to get by and jaywalkers skirt through to take advantage of the frequent snarl. Fed-eral Express, UPS, and Brinks trucks stop to make frequent pickups and drop-offs. Cars blare impatiently as trucks unload.

This one-block hive of industrious activity and spectacle extends along West 47th Street between 5th and 6th Avenues. This block holds the great-est concentration of offices and retail establishments in the district, though the district extends from approximately 45th to 52nd Streets and includes parts of the blocks along and east of 5th Avenue. Off 47th Street, other of-fice buildings provide wider, more spacious entrances. Shiny floors, banks of elevators, and an array of security clerks perched high greet the lost tourist, the private customer, the diamond dealer scurrying to appointments. In be-tween are other businesses—the Gap, banks catering to the jewelry trade, pharmacies, trade stores keyed to the jewelry business, Barnes & Noble. The famed Gotham Book Mart, literary home of poets and writers through-out much of the past century, is nestled among the jewelry offices and store-fronts on 47th.

The front layer of the small booths on 47th Street provides window dis-plays of goods. Scores of offices, most of which do not deal directly with the retail trade or with the private customer are crammed together above street level. In these offices people trade rough and polished goods with one an-other. Here too are the small factories where the rough goods are trans-formed into the polished goods that will then be set into rings, watches, brooches, necklaces, bracelets, and pendants. The finished pieces will then be sold directly to individual customers ("privates") and/or find their way to jewelry shops. The offices run the gamut from elaborate and sumptuous to spare and unadorned.

Diamonds are a Rorschach test of sorts. The mention of them elicits a range of responses. Their beauty, their exclusivity, their monetary worth con-vey a host of connotations and are therefore symbolically rich. Geologically splendid items, diamonds are symbols of colonialism and repression, the means of survival for persecuted peoples fleeing certain death, the way rebels finance wars, the stuff of lurid murder mysteries, the method of money-laun-dering schemes. The district provided an arcane background for an episode of *NYPD Blue* a few years back as well as for the Dustin Hoffman-Laurence Olivier film, *Marathon Man*. Diamonds periodically figure prominently in the news. Civil wars and corrupt leaders in Africa continue to be financed by diamond smugglers buying a steady stream of arms and ammunition.

One of many secured armored trucks on 47th Street. (Photo by author)

Other implications surround 47th Street as well. "Oh, how exotic!" and "It's completely Hasidic, right?" were some responses I received from outsiders when they heard of the research. Others remarked that the street was either crime-ridden or safe and wondered about the influence of organized crime. They urgently lent me their diamond romance or murder mysteries.

Others, both outside and inside the business, said, "Schlock!" when I asked what they thought of the reputation of the street. People inside the business told me I had to distinguish the businesses on street level with them upstairs. As one manufacturer told me: The guys who sell crap on the street, they're the ones who give 47th Street a bad name; we don't deal with them.

Sometimes the traders would explain that they meant one ethnic group rather than another was responsible for whatever bad reputation 47th Street carried. "Those Russians are so dishonest!" was one protest. Others complained about Israelis; still others, the Hasidim. Groups in close contact, brushing up against one another, vying, disdaining, and collaborating, maintaining their prejudices, and making exceptions for the individuals with whom they worked.

I was mostly going upstairs, above the street and all its clamor and rumors and intrigues to mill around with the wholesale traders and manufacturers, ethnically diverse, religiously varied, mostly male though resolutely peppered with women, and both young and old. These are the people in the

Famous Gotham Book Mart nestled among jewelry firms on the street. (Photo by author)

smaller, generally less prosperous firms who are far removed from the hype of the business though they are entwined with and affected by events and circumstances involving diamonds worldwide. I sought entry into this quieter, older world quite separate from the street noise yet very much of the modern world—and street-wise as well.

The entryway to the Diamond Dealers Club on West 47th Street is somewhat dim and narrow, but it is lined with large mirrors on both sides which reflect the available light. Groups of men walk to the back of the hall and stand assembled in clumps waiting for the elevator. A man sits on a perch behind a boxed pedestal in the corner at this end of the hallway. Dressed in gray, he regards each person with an almost imperceptible look, then pauses his gaze on me and asks blandly: Can I help you?

The security mirrors above him are trained on all of us, and I explain that my uncle is waiting for me upstairs. He asks his name and calls upstairs on the phone. Yes, it's all right to go up. The elevator has arrived and the group of men enters it. I press into the elevator with them and watch our reflection in the mirror on the opposing wall until the doors close. The men register my presence and ignore me. They greet one another with little nods, exchange a few words. A few speak together in Yiddish, Hebrew, English, Flemish. Several are dressed in long, black caftans and black hats. Some of the men wear long, white beards. Others wear *yarmelkes*, the only visible sign of their Orthodoxy. Some, in contrast, look like typical American businessmen, wearing business suits, clean-shaven, grasping sleek briefcases.

The doors open onto the tenth floor and we enter the gray foyer of the Diamond Dealers Club. Behind the glass windows ahead of us is the club itself. Two or more security men dressed in gray uniforms who sit on the other side of the glass look and nod at the men swiping their security cards through the two turnstiles on the right and then through the locked door into the club. Some greet each other ceremoniously with handshakes and animated talk. The members of the club pass through, and when it's my turn, I self-consciously nod at one of the security men. They signal OK since they've heard from downstairs, and I am let through the turnstile and then through the door. Usually my uncle Moishe greets me. He strides briskly toward me, dressed nattily in trim brown suit and tan tie, his clean-shaven face a young-looking contrast to the sea of bearded men swarming around. C'mere, Renée, he says, I want you to meet someone. He takes me by the arm.

◆

The gemstones glitter chaotically in a necklace, kaleidoscopically reflecting the light's zany colors. A surgical cutting tool embedded with a thin slice of industrial diamond slides through the patient's skin with lithe precision.

The golf club's sweet spot is hardened just right by a perfectly placed film of diamond. These three products and many others are the result of a long process. It starts with the ancient and massive rock formations that are forced to slowly yield to the mining that laboriously pulls the rough pebbles from their sludgy matrix. Then begins the countless manipulations of political, financial, and individual interventions that finesse and fuss over the stones.

People do these jobs. Perhaps a few million of them line up along the chain of activity from mine to customer in this $7 billion global market of uncut stones[1].

They include the freelance diamond prospectors, the geologists, geophysicists, surveyors, and others involved in scoping out new potentials for mining; the mine owners, operators, and diggers involved in extracting the raw diamond material from deep inside the earth, on beaches, in riverbeds, and recently, from the ocean floors; the bankers, investors, and financial analysts; the diamond handlers who sort and classify; the diamond manufacturers for both gem and industrial diamonds, who go by names like cleavers, sawyers, bruters, cutters, faceters, and brillianteers; and the government officials and regulators from agencies like the Internal Revenue Service and the Federal Trade Commission at each country of origin, at each border of trade, and at almost each sale along the line. At various points in the process the gemologists inspect, rate flaws, uncover synthetic modifications, grade and certify the stones and deem value to them. Chemists and other scientists invent procedures and fillings to improve, synthesize, and detect artificially treated diamonds. Brokers and dealers buy and distribute diamonds at each point in the process. Each has a vital role in the flow of this sprawling trade.

Add the people who populate the large and small diamond manufacturing and selling companies; the small jewelers in little, elegant stores on main streets and malls; the large retail chains, such as Kmart and Zales, that dominate suburbia and spread inexpensive diamonds to middle America; and the elite kings, such as Harry Winston and Tiffany's in their magnificent storefronts on 5th Avenue, who provide actresses, princesses, and moguls with their tiaras and rings. De Beers jewelry stores will eventually be among the mix.

The diamond manufacturing and distributing cartel, De Beers Consolidated Mines, and its Diamond Trading Company (DTC) with offices in London, Johannesburg, Lucerne, and elsewhere employ thousands of people, including financial analysts, gemologists, and detectives. The trade also includes smugglers, money launderers, guerrillas, and local despots, who sit in their clandestine bungalows and makeshift palaces, roam the borders and interstices within and around the tight but porous net of the international

---

[1] The diamond retail trade is estimated to be about $50 billion, about half of which is in the United States (Cowell 2001).

trade, and skirt and create danger to wrest inordinate profits from the minuscule stones.

Individual dealers, brokers, and manufacturers trade in the local arena of their cities, but many of them are national and international players too. Favorite big buyers of De Beers fly to London to receive their goods that they in turn distribute to other intermediaries, dealers, and manufacturers. Others fly to Antwerp and elsewhere to be second in line after the De Beers buyers and to capitalize on the goods that flow on the open market outside the reach of the giant cartel. Traders coming through a major city's diamond club are often en route to jewelry shows in Las Vegas, other clubs in other cities, or auctions in Geneva or New York. They may have offices in Florida, Germany, Brussels, or Texas. Buyers from Singapore, Bombay, Tokyo, and Tel Aviv appear frequently in the New York club to purchase goods for their customers and factories at home.

People in the diamond trade are not only local but also very global. Nowhere is this more true than in New York's diamond district, where the trade is more than 95 percent Jewish. In this labor of quintessentially nonessential goods, Jews have used this trade to stay separate from and vital to the economies of the societies in which they have lived.

◆

Some background first.[2] What is the stone that emerges from the ground opaque and cloudy, rough, and common? Diamonds are not rare; in fact, they are found throughout the world. Their quality varies widely. Yet people have died for them, their conquests have spawned magnificent legends and great literature, and they embellish their owners with prestige. The word diamond derives from the Greek *adamas,* meaning unconquerable.

Diamonds are not a mystery. They are created from heat and carbon compressed in deeply recessed volcanic magmas millions of years old that come to the surface through eruptions and soil erosion. The diamond features a unique crystal arrangement of carbon atoms surrounded by four neighboring carbons with a strong covalent bond that distinguishes it from other carbon minerals, such as graphite, and gives the diamond its singular properties. As the hardest known substance it is the most resistant to scratching and the best material with which to scratch other surfaces. Its high density (the carbon atoms are closely packed) and stiff structure slow the velocity of light traveling through it and make it extremely durable. It is thermally but not electrically conductive, making it well suited to electronic applications. As the light interacts with the electrons in the diamond, the

[2] A fine source on the physical origins and traits of diamonds is Harlow 1998a, especially articles by Kirkley; Levinson; Harlow, Shatsky, and Sobolev; Fritsch; and Harlow 1998c. Also see Krajick 2001 and Hart 2001. For the cosmic connection, see Benedetti et al. 1999, Leary 1999.

light is bent and produces a high refractive index that gives diamonds their high reflectance, or the amount of light reflected off the surface. The intense glittering effect is created when diamonds are cut to maximally exploit these physical properties.

Mined diamonds are as much as three billion years old. Extracting diamonds from the ground is laborious and dangerous. Diamonds are usually mined from kimberlite or lamproite volcanic rock deposits. Crystallized 200 kilometers or more below the earth's crust in the upper mantle, diamonds travel to the surface through volcanic pipes. When the pipes reach the surface of the earth, erosion and other weathering release them. Diamonds may stay fairly close to where they emerged from the earth or travel far via rivers to alluvial deposits. They also appear in onshore and offshore marine deposits. Gems that surfaced nearly three billion years ago in South Africa now wash ashore at closely guarded Namibian beaches. As a writer for the *New York Times* noted, "As the volcanoes that spawned them eroded, the diamonds were washed hundreds of miles down the Vaal and Orange Rivers, and spread up the coast by powerful tides. The river trip cracked the flaws, leaving gem-quality stones; and the river is a natural sorter" (McNeil 1998a).

Diamonds have also been formed outside the earth's mantle by more unusual means: perhaps a "collision between two tectonic plates [that] drove a slab of cold continental crust to . . . depths [of 120 km] . . . became detached from its down-going neighbors and was virtually squirted back toward the surface along a slippery boundary, aided by its own buoyancy" (Harlow, Shatsky, and Sobolev 1998:66–67), or even a cosmic collision, such as the impact of an asteroid with the earth, changed carbon into diamonds. Diamonds may also be formed from electrical discharges to methane and other gases from dying stars. Gaseous outer planets like Uranus and Neptune may even have diamonds near their cores because of their methane-rich gases.

Most of the color in diamonds is the result of impurities or defects in the atomic structure. A pure, perfect diamond is extremely rare and is colorless because no visible light is absorbed. Very slight coloration is generally undetectable by unaided eyes. When impurities alter the incoming light, they eliminate some colors of the spectrum, and we perceive the remaining colors. As impurities enter and new layers of diamond are added, subtle variations in the coloration of the stones are created. How we see the color of a diamond is also influenced by aspects of luminescence (the emission of visible light by materials subjected to some kind of excitation like radiation).

Nitrogen, the most common impurity, makes a diamond look yellow. Boron produces a blue color. Common brown diamonds are usually caused by a poorly understood process called "graining" (thin parallel "lamellae," or plates, that appear brown) due to a structural deformation at the atomic level. Red, pink, and the very rare purple diamonds derive their color from pink graining and deformation of the diamond structure, not from manganese as formerly thought. Rare green diamonds seem to owe their color to a thin superficial layer that has become naturally irradiated. The rarest color, orange with no trace of brown, is thought to be due partially to nitrogen. Violet, black, white, and gray are other colors found in diamonds. Because the cause of some of the colors remains unknown, the mystery and appeal of these diamonds has been considerable and helps explain why people have sought to add or to alter color in diamonds through the years using clumsy as well as sophisticated means. Pink nail polish was used to alter the color of a diamond at a modern auction, and other scams proliferate (Callahan 1996).

Strongly colored diamonds are rare (approximately 10,000:1) and are currently more popular and valuable than faint colors, echoing the preference of certain dynasties in the past, including the Mogul rulers of India, the Russian czars, and the English kings.

Different mines in the world produce different colored diamonds. Yellow diamonds (formerly called "cape" diamonds because they were often found in the Cape Province of South Africa) are the second most common colored diamonds after brown. Small amounts of pink diamonds have been found for centuries in India, Brazil, Indonesia, Tanzania, and now more regularly in Australia's Argyle mine. Red diamonds have been found in Brazil, Borneo, and Australia.

When diamond-bearing regions of the earth have been accessible to exploration, they have been or are being surveyed and exploited, though some treacherous areas still remain out of reach. Telltale minerals left behind by volcanic pipes (such as garnet and zircon) point the way to diamond discovery and exploration, and some recent geophysical technology may help uncover new diamond sources. Erosion, weathering, and river flows spread diamonds from these ancient pipes over such vast regions that the original sources of the diamonds are not always known.

Different mining processes are used for the specific type of diamond deposit—primary, alluvial, or marine—in question. In primary mining a large pit with shafts and tunnels underneath is excavated in an area of pipe rock determined to have good potential until hard rock is reached. Material is extracted by blasting and is then crushed to remove the diamonds. Small diamonds can be lost in this process and unless spotted soon enough, even larger ones can be crushed too. A diamond manufacturer told me: "With mining, a lot of stuff doesn't get handled by people. It gets handled by machinery and finally it gets handled by people doing the grading. They need to dig out many tons of ore to get to one little diamond. So it would be prohibitively expensive to do it by hand. They go into a machine that crushes the rock and the largest stones [diamonds] probably get broken up very often. Once in a while they found [a large stone]. Today, probably the [large and famous] Cullinan diamond probably would have gotten chewed up."

Because diamonds are heavy, they sink in water suspensions, then are extracted by a grease belt that adheres to the diamonds or by an x-ray fluorescence separator that activates a photo detector that in turn triggers a jet of air that hits the diamond into a collector box.

Alluvial mining was the first major way of finding diamonds, but because it produces a smaller yield it is used only by individuals and small operations. When diamond-bearing streams are located, the ground is dug up and the diamonds are extracted. Using a sieve and a grease table, searchers sort through the concentrate. Because these diamonds are easy to retrieve, they are the primary means guerrillas have used to finance the recent wars in West African nations, including Sierra Leone.

Marine mining has enormous potential though deep waters require expensive and extensive technology. Though the force of the ocean destroys stones of inferior quality, better ones endure. The largest known areas of gem-quality diamonds are along the west coast of southern Africa. Vast diamond-strewn areas that are the products of thousands of kimberlite pipes become buried and encrusted. Mining takes place in some of the crevices that the waves have cut over the millennia. Some diamond hunters search beaches on knees or dive in shallower areas off the coast.

◆

Once out of the rock, stream, or ocean, diamonds are determined to be either of industrial or gem quality and are grouped by size, shape, quality, and color. Gemstones are polished into shapes such as emerald, pear, marquise, oval, triangle, and the most common, round brilliant. Current diamond cutting techniques were developed long ago and were described by Benvenuto Cellini in the sixteenth century; "only the mechanisms, refinements, and degrees of automation have changed" (Harlow 1998c:231). A rough diamond destined to become a gem is examined with a loupe to decide how it will be cut, taking into account the location of the stone's imperfections (inclusions). The cutter tries to position the imperfections on the outside so they are easily removable. He or she wants to produce certain angles inside the stone that will reflect the most light from the top and create other angles that will separate colors the best. Achieving both perfectly is impossible. The cutter also has to create the most pleasing shape with minimal waste. He or she marks in India ink where the cutting will take place on the surface of the stone. A manufacturer explained the theory to me as follows:

"Light is coming in and every little facet acts as a mirror. It's pretty dull before it's cut. We call the surface the skin. The surface is not smooth and is sometimes completely frosted like sandblasting, and the light doesn't go in, so it doesn't come out. It gets diffused, but once you have facets on it, the light goes in and it gets bent, just like when you put a stick in the water it looks bent, and that's because of the specific density. The light comes in like this [he draws a diamond and lines of light on a paper], and it also comes in like this and then it gets bent and it goes through the stone here. Now here it acts as a mirror so at the same angle of incidence it comes [out] here. Here it comes out like this again, and every one [light ray] comes out. And the more angles you have and every facet is a little bit of a different angle. So you get all this light coming back out. That's what makes it. Once you make the facet, you have a mirror, and once you have a mirror, you have a possibility of refraction, right? Pure refraction, not iridescence.

"[Liveliness] is also on the angles on which you do it. Certain angles have more refraction than other angles. Now if you make the stone too deep on the bottom, it looks dark because the refraction goes out. When the light goes in like this and it comes out that way [he draws a long line], the light comes out on the side, so you don't see it. Aha! The life of the stone depends on the cut. This was figured out by some genius in 1910."

Another cutter explained how beauty is sacrificed when stones are cut along natural angles, resulting in a "too deep" cut and a greater weight (Blauer 2001). The cutting process involves four operations, known as cleaving, sawing and lasering, bruting or girdling, and polishing. First a notch is created in the diamond with another diamond, and then a cleaving blade separates the halves of the diamond along the grain line. (Sawing divides the diamond against the grain.) The center, or girdle, of the diamond is smoothed against a tool diamond held by a dop (a holder) on a circular lathe. Polishing each facet is accomplished by positioning the diamond in a dop on a wheel impregnated with oil and crushed diamonds. The diamond is checked frequently and is repositioned to polish all facets. The cutter's goals are to reduce imperfection, enhance brilliance and glitter, and create a beautiful shape, or make.

Polished gem diamonds are graded according to the "4 C's:" carat weight, color, clarity, and cut[3]. Increasingly, a fifth "C," the certificate (officially called a grading report) has been added to the trade. The color grade is determined according to an internationally used GIA scale that arrays diamonds from colorless to light yellow as D–Z (D equals colorless and Z light yellow). The more strongly colored diamonds are called fancy colors and have their own scales. Clarity grades describe the flawlessness of the diamond's internal composition, from internally flawless (IF) to internally included (e.g., VSI). The grading report also notes the proportion or make of the stone.

Non-gem diamonds have varied industrial and specialty uses for plastics, metals, and glass and are used in cutting tools, abrasives, grinding powders, and pastes. They are fitted in tiny dental drills and used in rock and earth deep drilling, in styluses for record players, and in the ceramics and stone

---

[3] The three C's are an important theme in the diamond business (Shigley and Moses 1998). Related to a diamond trader's reputation, an industry banker considered other C's: "Capital, Culpability, and Character—at our bank, we think that character is the most important C" (Bernstein 1992:154).

Hasidic cutter adjusts the stone on the dop. (Photo by author)

Diamond placement on the dop. (Photo by author)

industries to smooth surfaces. Their low friction and endurance makes them ideal in wires, fiber-optic cables, and surgical instruments since they slice but don't drag or deform skin. Space windows made of diamond are transparent in all wavelengths and can withstand great heat and pressure.

Attempts to create artificial diamonds by reproducing the high heat and pressure under which diamonds are naturally created have been made since the eighteenth century. In 1955, graphite was transformed into diamond using the pressure equivalent to the weight of the Eiffel Tower on a penny. Most synthetic diamonds are used like natural industrial-grade diamonds. Recent spectacular achievements in synthesizing diamond have occurred through chemical-vapor deposition (CVD). Methane and hydrogen gases are combined at high pressure to result in a deposit on a substrate of material such as silicon. Initial film deposits have given rise to thicker wafer-like deposits.

Gem diamond imitations range from inexpensive colorless materials, such as cubic zirconium, to synthesized diamonds. Until recently, there have always been ways to detect false from authentic stones. General Electric produced jewelry-quality synthetic diamonds in 1970 but at a prohibitive cost precluding production.

Natural diamonds can also be treated in various ways to improve or enhance their appearance, and these treatments can almost always be detected by the GIA. Treatments can alter the color and the clarity and can reduce the imperfections. Clarity can be enhanced by fracture-filling and by laser removal of inclusions, both of which require full disclosure by law and by the World Federation of Diamond Bourses (WFDB). While appearance may be enhanced, durability is threatened by these processes.

The actual and potential industrial applications for synthetic diamond are enormous, such as high-temperature electronic appliances that are too hot for silicon, improved flat-panel laptop computer screens, and spacecraft windows. CVD on golf clubs currently makes the sweet spot harder, and elongated diamonds polish contact lenses and engrave Steuben glass. The wafers of polycrystalline diamond that can be routinely produced and allow enormous thermal conductivity could vastly increase the speed of interconnections in computer applications. The CVD diamond is being explored to detect nuclear and ultraviolet radiation. In the future synthetic diamond could possibly be used in high-temperature active electronic devices and new TV screens. CVD-produced ultra-thin diamond tubing might be used in fiber-optic cable and for tiny hypodermic syringes in microsurgery. The potential for diamond application is exciting (see Hazen 1999 and Collins 1998).

◆

Why and how do things become imbued with value? The great irony of diamonds is that they continue to be valuable though they are no longer rare. This paradox is a core feature of the diamond business. Though the diamond is now mined on all continents but Europe, it maintains an aura of uniqueness and prestige. Its high value persists because of the successful marketing and distribution strategies of the global cartel, De Beers. While the diamond has become commonplace and generally rote-produced and the business increasingly commodified, the mystique is hyped by marketing, and the value stays high because De Beers has titrated its distribution to maintain keen demand and exclusive pricing.

Beyond these features, however, diamonds, like other things, can have social lives (Appadurai et al. 1986). Certain diamonds, named and well-known, have achieved fame because of their size, color, or clarity. Some fabled diamonds have the mysteries of legend and curse attached to them. Even though the marvelous detective Sergeant Cuff, who hunts the lost diamond from India in *The Moonstone,* remarks at one point, "I also think a rose much better worth looking at than a diamond" (Collins 1994:630 [1868]), the entire novel is taken up in pursuit and dread of this haunted stone. A wonderful joke I heard in the DDC pokes fun at the legend-making connected to diamonds: Mrs. Plotnick is finally willing to sell the famed Plotnick diamond. What a magnificent opportunity. It is huge and glorious with unsurpassed clarity for a stone its size. But there are murmurs about it: "Do you know about the curse?" they whisper. "No, what curse?" Pause. "Mr. Plotnick."

Like other things, diamonds can take on value, as well as gain and decrease in prestige. In addition, they have histories. As people can be commodities (the most extreme example is slavery), so things can be personal and unique (an Oriental rug, a painting, a diamond). As exchange creates value, politics and power determine the kinds of value they acquire. Kopytoff notes that a thing can have a "biography" and asks, "Where does the thing come from and who made it? What has been its career so far, and what do people consider to be an ideal career for such things? What are the recognized 'ages' or periods in the thing's 'life,' and what are the cultural markers for them?" (1986: 66–67).

Some of the meaning and value of certain diamonds lies in the uniqueness (or singularization) of certain diamonds. These unique things can be valued because of a sacred status attributed to them. On the other end, Kopytoff notes, commoditization "homogenizes value." Commoditized prestigious items, like most diamonds, can have "terminal" value when no further exchange is likely, even though a common selling strategy is that they are potentially resalable. Kopytoff refers to "the promise that oriental carpets, though bought for use, are a 'good investment,' or that certain expensive cars have a 'high resale value'" (1986:75). In the 1970s, diamond investment scams reflected an illegal version of this selling push.

Terms like singularization and commoditization are rare or nonexistent in the languages of diamond exchange on 47th Street, but the reality of these poles is clear. I heard traders note the mystery surrounding value as they highlighted what they found bizarre. During the Sotheby auction of Jackie Onassis's things in 1996, for example, they were perplexed and amused. One contrasted normally stable diamond prices with the stratospheric prices at the auction. If you're Jackie Onassis and "you have a *schmatte* [rag] that you blew your nose in, it goes for $75,000," said one trader. Here, you see, he instructed me, value is relative. If I have it, it's a *schmatte;* if Jackie has it, it's worth something!

Though New York diamond dealers understand commoditization—how diamond specks only as big as salt are mass-produced by computers and by children in Third World countries for thousands of identical bracelets and necklaces—they know how larger diamonds are valued for their uniqueness. In New York the big diamonds are more likely to be individually known—singularized.

◆

Small numbers of diamonds were found in Borneo beginning in the sixth century. From earliest times the special physical attributes of diamonds triggered the constellation of meanings that each culture assigned them. Noted for incredible hardness and flashing color, they were elaborated in India as God-given and as the essence of purity; legends touted these qualities (Harlow 1998b:119).[4] Diamond-studded Roman jewelry in the first centuries of the Common Era was believed to ward off evil. Associated with pagan rituals, diamonds found less favor with the early Christians. In the Middle Ages diamond talismen were thought to confer power on men. Marco Polo encountered diamonds in his travels. Diamonds could cure madness, nullify magnets, kill, prevent natural disasters, and enhance love, and their powder was ingested as medicine (Legrand 1980).

Before Jews became involved in diamond trading in the first millennium, diamonds were important to their rituals and beliefs. From the biblical era onward, abundant references to diamonds occur in religious texts such as

[4] Diamonds were mentioned in Indian manuscripts two thousand years ago. In an important Buddhist text from India dating from possibly the third century, the diamond was linked to religious virtue. The text admonishes the truth-seeker: ". . . Just so, O King, should the strenuous Bhikshu [truth-seeker] be earnest in effort, be perfectly pure in his means of livelihood . . . the first quality of the diamond he ought to have . . . just so, O King, should the strenuous Bhikshu, earnest in effort, never mix with wicked men as friends . . . the second quality of the diamond he ought to have . . . and again O King, just as the diamond is set together with the most costly gems; just so, O King, should the strenuous Bhikshu, earnest in effort, associate with those of the highest excellence . . . the third quality of the diamond he ought to have" (from the Questions of King Milanda, quoted in Harlow 1998b:121).

the Torah, Mishna, Midrash, and the Talmud. Special powers were attributed to diamonds, and wisdom was compared with them. Like a rare gem the Torah is "a precious jewel in your hands" and "cannot be redeemed and has no substitute" (quoted in Ilani et al., 1991). Anxiety over lost jewels is described in the Midrash, and the Halachic [legal] writings describe the adjudication of disputes over gems and diamonds. Gems adorned kings' crowns, thrones, Torahs, musical instruments, necklaces, rings, hats, scarves, and the fringes of clothing.

In the second century B.C.E., Plautus wrote of rings as tokens of love (Scarisbrick 1998:163–64), and diamonds were important in Shakespeare's and Marlowe's plays, as well as in works by Thomas Carlyle, Wilkie Collins, and Anthony Trollope. Numerous ornamental jewels with diamonds are found throughout Europe, India, and Russia, and portraits of nobles sporting diamond jewelry have been abundant since the Middle Ages in Europe.

The first major place that diamonds were mined was in the riverbeds of India in the seventeenth century. The scarcity of diamonds that reached Europe fed the idea that diamonds were for royalty only. Though there were few stones of actual gem quality (Levinson 1998:74), some of the most famous gem diamonds, such as the Koh-i-Noor, Great Mogul, Hope, and Nizam, came from India. Portuguese Sephardic Jews who traveled from Surat in India to the Portuguese Goa to Venice, the East, and Belgium dominated this early and tiny diamond trade.

The trade centers shifted over the years, initially concentrated in Venice and Bruges in the thirteenth century, then moving to Antwerp when Bruges's river became heavily silted in the early fifteenth century. Diamond cutting has been practiced in Antwerp at least since the fourteenth or fifteenth century. Trade in imitation gemstones was banned in Antwerp in 1447. When Jews were expelled from Spain and Portugal during the Spanish Inquisition of 1492, they took the small gems with them to Amsterdam. Though Jews in Holland did not have Dutch citizenship, they had religious freedom and protection. Lisbon became important after Vasco da Gama's voyage to the Indies. The trade between India and Europe increased.

Let me add a few words about traders. Members of minority groups are often stigmatized for their activity, and those Jews who trade, such as in the diamond industry, are no exception.[5] Ethnically distinct traders and other middlemen are often considered strangers to the society in which they live

---

[5] For sources related to Jews, diamonds, middlemen, and trade in general, see Gross 1975; Ilani, Goldberg, and Weinberger 1991; *Encyclopedia Judaica;* Kockelbergh et al. 1992; Laan 1965, 1975; Shor 1993; Gutwirth 1968; Legrand 1980; Kanfer 1993a; Federman 1985; Lipschitz 1990; Yogev 1978; Liberman 1935; Balfour 1992; Kertzer 2001; and Zenner 1991.

because of their economic ties to other societies and their cohesiveness and difference from the masses. Standing at the borders between groups, they occupy a special no-man's land, a cultural liminality. Whether moneylenders, tax collectors, or diamond traders, they often derived a privileged status from their closeness to the powerful, but that status was precarious and often carried stigma, as well. Entrusted with secrets, they were likely to be scapegoated. "Middlemen minorities—which are often penalized minorities—also combine the paradox of relative economic success and even affluence with ultimate powerlessness which permits their victimization," Zenner says (1991:xiii).[6]

As most Indian stones were imported to Europe through London, London became the center for rough diamonds while Amsterdam was the cutting center. In 1618, Jews were driven from Eastern Europe, and many went to Amsterdam. The business was difficult, subject to large economic fluctuations and fierce rivalry between Portugal and Holland. Working conditions were harsh, and poor ventilation helped make tuberculosis an occupational hazard. Training required a six-year apprenticeship.

Because London controlled the Indian gems, many Sephardic Jews from Amsterdam resettled to England in the seventeenth century and received special permission to import diamonds from India. Records of Portuguese Jewish diamond polishers date from 1615. After 1700 prominent Ashkenazi Jewish merchants imported rough diamonds and exported silver and coral back to India. Records note several Jewish diamond agents in India in 1750.

But Jews did not have full citizenship rights in England, and, as they were unable to become naturalized citizens, they had to pay alien duties. The anti-Jewish attitude of Christian merchants made business difficult. Yogev writes, "The weight of a tradition which had moulded Jewish economic life for generations, the limited capital resources at the disposal of the majority

---

[6] Nineteenth century German sociologists theorized about the antagonism successful Jewish businessmen engendered. Weber suggested, for example, that their "pariah" status was the result of a "this-worldliness" that made them opportunistic. Kertzer (2001) demonstrates the pervasive anti-Jewish attitude in Europe promulgated by the church for hundreds of years. Bonacich (1973) noted that when U.S. minorities work for low wages in jobs others avoid, other groups react against them. Abner Cohen (1974) examined how groups create networks of help and stay inconspicuous at the same time.

Jews, in addition, carried extra burdens. They were routinely considered the cause of many of society's ills. Throughout Europe they often were forced to live in separate locked-in areas (the term, ghetto, was coined in Italy). While the church forbade Christians from what it deemed the usurious job of lending money, for example, this vilified position was one of the only roles Jews were permitted. They were often considered enemies of the state and untrustworthy foreigners—not merely members of a separate religion. Practicing a different faith from the mainstream, resisting conversion into Christianity, Jews were also regularly blamed for killing Christ, of committing ritual murder of children, and other outrageous charges by church doctrine and officially sanctioned pronouncements (for the role of the Vatican, see Kertzer 2001).

of Jewish merchants, their dissociation from the land, their still inferior legal status and the jealousy of the Christian merchants, all prevented the Jews from making full use of the new economic opportunities which England offered" (1978:19).

Antwerp benefited from the fact that the finest stones from London were cut in Amsterdam. Because Antwerp cutters received the rejected, inferior stones and learned to cut them to maximize their value, they became highly skilled and developed a superior reputation. Later attempts to expel Jews from the city were resisted because Jews had become too important to the city.

Jews who were attached to German and other courts provided important links for royal distribution.

Jews remained prominent in the trade after diamonds were discovered in the Portuguese colony of Brazil in 1730, particularly in the cutting industries of Amsterdam and Antwerp. Continuing their middleman role, some of these diamond workers were connected both to the financial houses and to the upper classes that bought the jewels. As huge numbers of smaller, more inferior diamonds began to flood Europe over the next decades, prices plummeted, and non-royalty began to acquire them. The trade increased with the new finds, but the flip side was lower prices and greater disorganization. While the Rothschild and other international banks helped control the flow from Brazil, *garimpeiros* [independent prospectors] resisted the imposition of order, and they smuggled diamonds at low prices.

Though little has been written about women in the diamond business, their labor was apparent but restricted during this time. Kockelbergh, Vleeschdrager, and Walgrave state that after 1722, women were allowed to be cleavers. The guild called the Nation of the Diamond and Ruby Processors permitted women to help the "master" turn the wheel and made an exception for widows, who were allowed to carry on their husbands' shops. "Like the customary law in other artisans, the Nation did allow the master's widow to continue to run the diamond polishing workshop in order to secure an income for her family. She could keep her trainees which her husband had recruited, but could not take on new ones. During times when the diamond trade was thriving, the masters were permitted to teach their wives the skill of diamond polishing" (1992:93).

From at least the eighteenth century, women were needed to assist and were feared as competitors. In 1733, the Nation restricted what Kockelbergh et al. call "clandestine women laborers" because they were very good polishers, so good as to threaten the masters.

In Jacobs and Chatrian's monographs from the 1880s, from which Kockelbergh et al. derived much of their material, the "unquestionably high competence of women" was discussed (1880). They praised women's skills, refinement and dedication and unblinkingly noted their lower wages and longer hours. Lipschitz's history of the Amsterdam trade observed that women worked when families polished diamonds in their own homes in the eighteenth century. "Here women, who provided cheap labour, worked the polishing machines. When the diamond trade slackened [as the Brazilian diamonds dried up] and even cheaper means were sought, the women were replaced by horses," she noted sardonically. But nonetheless, when steam power replaced horses, women were valued as good polishers because they were delicate (1990:33).

However, it seems that the status of women's work remained a thorny question. In 1894, the International Congress of Diamond Polishers debated whether to abolish female labor—widows of diamond polishers, those who assisted at the wheel, and independent cleavers—from the diamond workforce altogether (Kockelbergh et al. 1992:138).

During the last decades of the nineteenth century, exploration, colonization, and trade increased in the world. Gold rushes occurred in California and Australia. Yiddish-speaking Jews throughout Europe, primarily working in middleman roles, were often restricted in ghettos, and in general, subjected to discrimination, legal constraints, pogroms, boycotts, and expulsions over the next century.

Then diamonds were discovered in South Africa in the late nineteenth century, producing an oversupply in Europe. The flood of diamonds from Africa invigorated the cutting industry in Antwerp and Amsterdam but depressed prices. Diamond clubs with uniform and orderly rules were established in these cities and supplanted cafés as trading places.

Though some reports attribute the Bushmen using diamonds earlier (Balfour 1992), diamonds were officially discovered in 1867 in South Africa when an eight-year-old boy reputedly picked up a large pebble that was later identified as a 21-carat diamond.

The South African story deserves special attention.[7] For the first time diamonds were mined by creating open pits into the kimberlite pipes. The kimberlite mines supplied about two to three million carats per year and 95 percent of the world market supply between 1872 and 1903 (Levinson 1998:83). Gazing at the large open pit of the "Big Hole" at Kimberley, Anthony Trollope felt "as though you were looking down into a vast bowl, the sides of which are smooth as should be the sides of a bowl, while round the bottom are various incrustations among which ants are working with all the unusual energy of the ant-tribe" (quoted in Green 1981:37).

---

[7] Histories of the South African diamond story are plentiful. Some of them are: Hahn 1956; Worger 1987; Roberts 1972; Gregory 1962; Newbury 1989; Kanfer 1993b; Jessup 1979; and Turrell 1987.

Complicated relationships among classes of owners and workers developed over the next decades. Europeans challenged the authority of black Africans who owned property in Kimberley, and unfair working conditions (low wages, insufficient food, and unsteady work) helped foster illicit diamond buying, which in turn led to repressive methods of control of the workers (Turrell 1987).

The diamond rush encouraged miners to cooperate as well as to compete, possibly laying the groundwork of the future cartel organization of diamond production and distribution (Spar 1994). Approximately 1600 people held separate claims during the Kimberley diamond rush (Levinson 1998:83). As claims proliferated and the dangers of open-pit mining increased, people began to amalgamate claims and to cooperate to create safer and more profitable conditions.

When prices dropped and mining problems increased, eighteen-year-old Cecil Rhodes followed his older brother to South Africa from England in the early 1870s and tackled both problems. First he devised a method of pumping the water from frequently flooded pit mines. Then he decided that control over production and selling would ensure the continued perception of the diamond as precious. After vanquishing competitors such as Barney Barnato for control of mining claims, he formed the De Beers Mining Co. in 1881. When he merged the company into De Beers Consolidated Mines in 1888, Rhodes controlled most of the world's diamond production. He established the foundation of the diamond cartel that Ernest Oppenheimer perfected. Rhodes also established Rhodes scholarships in his will and was the namesake of colonial Rhodesia. A mystery to many, this mercurial, sickly, and extremely powerful man died at age 48 in 1902.

Oppenheimer, the son of a German Jewish merchant (he later converted to Christianity after several personal tragedies), was born in 1880 and arrived in South Africa in 1901 by way of London. Encouraged by his father to go to England because there were few jobs for Jews in Germany, he and his brother found positions there sorting diamonds. Soon after landing in South Africa, he acquired the Premier mine and challenged the De Beers monopoly. After World War I Oppenheimer received backing from J. P. Morgan and formed the Anglo-American Corporation of South Africa. While Oppenheimer began to mine gold, he acquired a seat on the De Beers board by allying with Barney Barnato's nephew, Solly Joel, and acquiring enough De Beers shares to be appointed.

Meanwhile, immigration of Eastern European Jewish diamond workers to Western Europe tripled, with the number of Jews in Antwerp increasing from approximately 8,000 in 1900 to 25,000 in 1913. Though a tuberculosis epidemic in 1905 claimed many in the diamond trade, there were 10,000 diamond workers in Amsterdam by 1919. They began to unionize. New diamonds of smaller and more inferior quality were discovered in Belgian's colony, Congo, and Antwerp cut them. Other great new diamond finds outside South Africa increased the supply while demand plunged because of the Depression.

To keep diamond prices high, Oppenheimer established The Diamond Corporation in 1930 and created the Central Selling Organization in 1934 to be the central storage mechanism and clearinghouse by which diamonds throughout most of the world have been distributed.

◆

How Oppenheimer structured De Beers and the CSO proved pivotal to diamond trading. Rhodes's brainstorm, to control production and selling in order to maintain scarcity and to stabilize the market, entailed two steps: consolidating all the mines under one company and controlling selling through one channel. The DTC now controls the distribution.

In New York, De Beers and the CSO were benignly referred to as "the Syndicate." The cartel adds stability to the industry because the commodity's value is maintained as huge fluctuations in price are avoided. Diamond stability is in marked contrast to other gemstones such as emeralds, the value of which spikes and plummets drastically over time. In the single-channel distribution system, De Beers acts as buyer, and then it distributes and sells through the DTC. De Beers used Anglo for collateral when it had to borrow billions of dollars to shore up lower-price diamond years to keep prices and the market stable. The syndicate's ability to stockpile diamonds and to raise and lower prices according to inflation and demand has buffered the luxury trade from fluctuations in consumer discretionary spending.

When De Beers paid competitive prices to producers when prices were low, the cartel was not undercut, and De Beers became stronger in bad times. De Beers competed against independent traders to buy excess supply when markets were weak. This process was paired with controlling the supply so diamonds would not be overproduced and precipitate a drop in price. Plus, De Beers offered discretion to its producers: it didn't tell the world that Russia was collaborating with South Africa during apartheid, for example.

Terming cartels "a particular form of cooperation" (1994:2), Spar labeled the diamond cartel one of the most successful. In her view, cartel success relies on secrecy, exclusivity, and lack of formal rules—traits anathema to American values, not to mention the antitrust laws that outlaw monopolies. Though the United States has initiated several antitrust suits against De Beers since 1945—the latest in 1994—none have been successful thus far. De Beers and the CSO maintained the clout to keep producers loyal and threaten their viability if they defected. In the past various diamond entities attempted to operate outside the CSO, including the Mwadui diamond mine owned by John Williamson, a Canadian geologist in Tanganyika (now Tanzania), Israel in the late 1970s, and Zaire in 1981. Each, however, returned defeated to the fold.[8]

Those not in the CSO system risked an uncertain supply from independents who might not produce enough diamonds or who might join the CSO. The CSO could flood the market with stones of the same size and quality as those of a producer that wanted to operate independently. As large diamond supplies were being discovered and mined, producing companies negotiated contracts with the CSO or operated separately. Large independent producers could increase the percentage of goods circulating in the outside market. De Beers tried to stem this hemorrhage by buying stolen, smuggled, or otherwise "outside" goods. For example, a trade writer estimated that in one period as much as 10 million carats of diamonds (or 20 percent of world production) flowed from Congo to the market outside the CSO, and De Beers was the biggest buyer (Shor 1993:216). However, in the new millennium, the percentage of those operating outside the cartel has grown larger.

After De Beers bought, (now through the DTC) De Beers sold. The mechanics of distribution is as follows: The DTC sorts, grades, and packages the diamonds into parcels that are sold or placed in storage, depending on the state of the worldwide market. At ten sales, called "sights," that occur throughout the year, the parcels of diamonds are sold to the DTC's favored buyers, who fly to London from around the world. These sight-holders make order requests to the DTC prior to the sights; the DTC then prepares the parcels by adjusting the requests by its appraisal of worldwide availability and demand. Whether the resulting parcels the sight-holders receive correspond to what has been requested, they conform to what the DTC wants to distribute at the time. After the buyers view the parcels—presented in shoe boxes—they buy the parcels in toto or refuse them. Refusal means forfeiting the right to continue as a sight-holder. The sight system of releasing a special mix of diamonds into the boxes has the effect of keeping the most desired stones scarce and the less desired stones circulating; at the

---

[8] See Knight and Stevenson 1986. The Williamson example may parallel the experience of other mines. Because the British government was pleased with how De Beers managed the cartel (and supplied the Allies with industrial diamonds at stable prices during World War II), the trilateral bargaining between Williamson, Tanganyika, and London resulted in De Beers getting what it wanted. The authors concluded: "No doubt the experience of each country is unique in various ways. Nevertheless, in almost all important cases, the producers of diamonds and their governments have perceived advantages in selling through the CSO and disadvantages in breaking away from it" (1986:445). Spar (1994) has a good account of the relationship between Israel and De Beers prior to the diamond collapse in 1981. In an attempt at a separate defection in 1981 Zaire tried to go it alone. After the market was flooded with stones of the similar size and quality as Zaire's (probably by De Beers), Zaire signed a contract with De Beers in 1983 with less favorable terms than the country had had previously.

same time other plentiful categories of stones are stockpiled, preventing a drop in their prices. In economically difficult times the flow remains even. When the prices of other commodities rise and fall crazily or collapse, diamonds thus stay strong.

The sights have a mystique. Sight-holders are privileged since they have been deemed worthy by the syndicate. Diamond lore speaks of the rewards and punishments meted out by De Beers reflected in the kinds of allocations received by particular sight-holders. A diamond dealer in his mid-80s conveyed the aura of power these events hold in his story of how he regained the lapsed family sight:

"That was an interesting experience, going to London right after the war, 1946. I had never had a sight before personally, but my family had before and they kind of stopped it, and then I wanted to continue a sight again. Finally they called me up all excited; 'You can go in' [to see the person at Lloyd's]. It was a little brownstone building, very discreet. We went down the street: there was an old man in an army uniform. I say old man; maybe he was 40 years old. To me he was an old man. He was standing in front of the door. I told him who I was, and he looked it up and knew I could go into the building. I went upstairs and there I was, sitting, waiting to get in, and everybody was dressed in a cutaway. You know what a cutaway is? It's a jacket you wear for daytime weddings, long swallowtails in the back with a vest and stiff shirt with an ascot collar. I was the only one in a business suit. I felt so out of place. I was casual. They all were wearing striped pants, of course. Derby hat. Umbrella. And everybody was sitting there that way. The uniform. And finally I got in to see him. He was very, very gracious, very nice. 'Oh, I knew your grandfather well; I knew your family.' I explained how we needed to get sights, and he said, 'We'll look into it.' I didn't hear from him for a long time, but I did get a sight."

The cartel endures with a unique combination of controlling production, dominating the trade, and enticing customers with effective advertising. One writer offered the justification that the diamond is "a product extremely laborious to unearth, and with highly specialized uses, which the world has come to be persuaded to buy in steadily increasing quantities and at sensible and relatively stable prices. When times are hard, the jewellery trade is among the first to suffer. Therefore, stockpiling is of the essence, and its cost has to be shared throughout the industry" (Oliver 1990).

Ernest Oppenheimer's son Harry ran the companies from 1957 to 1994 and was a member of Parliament from 1948 to 1957. In 2001 grandson Nicky made drastic changes in the structure of De Beers.

◆

During World War I, Jews in Belgium who had emigrated from countries such as Germany, Austria, and Turkey were made to leave (since they were considered German nationals). Some went east and some went to the Netherlands. They were invited back to Belgium after the war and many returned, joined by increasing numbers of immigrants from Poland, Austria, and Hungary. During the Depression diamonds dried up and the prices plummeted. What demand remained for diamonds shifted to smaller stones. Once again, Antwerp's forté was needed.

As the industry began to recover from the Depression, the more penetrating and total disaster of Nazism and World War II destroyed a huge portion of the European diamond community. Perhaps 80 percent to 90 percent of the industry in Europe was Jewish at that time. The Germans raided the bourses in Amsterdam and Antwerp and took them over. Those Jews who escaped went to the United States, Palestine, South America, South Africa, Cuba, and Puerto Rico. When they had them, Jews attempted to use diamonds and other jewels to save their lives. One wrote, "The next morning we were 'liberated' by the German army. I left my parents-in-law unconscious but still alive. I remember putting some jewelry in their pockets just in case they made it, so they'd have something. And I had to run through the city of Calais, carrying my daughter, with the bullets flying over our heads. . . ." (Branch 1998:152).

Leon, a diamond trader, told me his immigration story. He came to New York from Antwerp in January 1938, to expand his father's diamond business. Leon's family came over later that year. Some of them had been born in Holland and some in Poland. They managed to find their way from Antwerp to Portugal, then got visas for New York and Curaçao, a Dutch colony in the Caribbean. Fourteen others then joined the ten of them. The safest way was to get to the south of France going along the coast. With their "merchandise" they bought a bus in which to travel together, not realizing that no one could drive it. Then one of the boys figured out how to drive it. When they ran out of gas, a pretty girl among them flirted with a gas station owner to give them gas. They put the merchandise in the hollow struts of the bus, knowing that the French would confiscate it if it were discovered at the borders. The boy drove through the Pyrenees at night, terrified. The mother didn't have a visa for Curaçao because she wasn't a Dutch citizen. Leon spoke to the Dutch consul in New York, who had good advice: write to the consul in Curaçao, notorious to refugees as an anti-Semite. Explain their problem, that they had visas for Curaçao but their mother was not a Dutch citizen. Ask for advice: would they be admitted after making the long trip there? As they expected, the Curaçao consul wrote that the mother would not be admitted and they would be refused. With written proof that Curaçao would refuse them, the family could now go to the U.S. consul, explain that they were stranded, and ask the United States to admit them. This is what happened.

During the war the loyalty of De Beers to its own interests was paramount. When the British and U.S. governments needed industrial diamonds to manufacture war material, De Beers initially refused to sell (perhaps fearing a drop in prices if the government didn't need the quantity it was ordering) and then overcharged them (leading to a federal antitrust suit in 1945). England pressured De Beers when President Roosevelt threatened to withhold attack planes. The Office of Strategic Services (OSS) charged that De Beers was meanwhile selling industrial diamonds to Germany. Antitrust investigations ended in 1976, when De Beers paid $40,000 and pleaded no contest (Jessup 1979).

After the war, Belgium tried to entice diamond dealers to return, but thousands stayed where they had settled. Whereas 55,000 Jews had lived in Antwerp prior to the war, only 10,000 Jews lived there in 1966. Two-thirds of the Dutch diamond cutters and dealers perished in World War II, and most of the survivors were older than 65, making it difficult to train a new generation. Fledgling cutting industries in Puerto Rico, Cuba, the United States, and Palestine began to take off during the war. Renewing its cutting industry started by the Jains before the seventeenth century, India sent dealers to set up offices in Antwerp, Palestine, and the United States. Under Nehru, India decided to concentrate on cutting smaller, less fine stones for export as Antwerp was decreasing its activity in these smaller stones. India gained its first sight in 1962, and by the 1980s approximately ten million Indians were employed in the industry (Patwa 1989, McDonald 1993, Westwood 2000).

Heralding the glamour and elegance of diamond jewelry to the public through ingenious marketing was magic. The massively successful "A Diamond Is Forever" ad campaign, launched in 1947 by the N.W. Ayer Company, was a perfect post-war pitch. It helped dispel drab vestiges of the war and put the notion of diamonds into the imagination of the American middle class learning to link love with the gem. Diamonds were becoming a "highly gendered story" (Westwood 2000). Hollywood movies powerfully melded fame, luxury, and glamour as major diamond companies gave movie stars glitter to wear onscreen. When a movie star popped the question with a diamond ring, big sales rang up in middle America. Diamonds were prominent in *Diamonds Are a Girl's Best Friend, Cat on a Hot Tin Roof,* and the James Bond movies, among many others. As much as movies helped diamonds sell, diamonds helped the movie business, too. The careers of Marilyn Monroe, Mae West, Greta Garbo, and Grace Kelly, to name only a few, soared as they were lavishly draped in the jewels. And to see Gwyneth Paltrow, garnished by a Harry Winston-lent diamond choker, elegantly accept her Oscar in 1999 was to recognize that the winning combination of gem and stardom continued happily (see also Spiegel 1998).

The 1950s were prosperous years, but smuggling continued to torment

the official trade. Individuals and groups had defied the established param-
eters of the business, capitalizing on financial and governmental instability,
and such traffic had proved impossible to staunch completely. Adventure in
the rich industry was enticing, especially when astonishing profits could be
plucked from the easily hidden gems. The stories I heard focused on the
1950s and stressed the adventure, liberation, and braggadocio of such men
(women may not have participated in such exploits) in those days. Tellers of
adventures enjoyed the trickster aspect of these tales. If a person was clever
in his audacity—could thumb his nose at imminent danger, or could survive
by wit or fluke—the tales were fun to relate and satisfying to hear. In a business so elaborately dominated by the big cartel, which was willing to spend
millions to protect its domain, there was pleasure in hearing of the fabled
men who had skirted the control of De Beers and others. Here was one:

"One guy, Maury, was in some godforsaken country, let's say British
Guyana, a real wildcatter in an area where there was a kingpin who had the
whole supply sewed up. And one day Maury found out that this kingpin had
offered $50,000 for his head. He didn't care for the competition. So Maury
decided this was not the place for him. Before he left, he called the kingpin
and said, 'I hear you put a price of $50,000 on my head. You know what?
Give me the $50,000 and I'll leave.' I don't think he got the $50,000. He
called from the airport. If you look at this guy's face—I heard this story sitting at lunch in the bourse in Antwerp—when I looked at this guy's face, I
could believe it. I just love that story."

De Beers has traditionally swooped upon diamonds outside its network
to buy them up so they could be stockpiled and held against diamond supplies that threaten to depress the market. A diamond manufacturer told me
a smuggling story that focused on getting the best of De Beers:

"In the '50s [in Sierra Leone] all you'd have to do is invest in a shovel and
start digging and you'd be liable to find [alluvial] diamonds, and [people
would] smuggle them over to Liberia [to] Arab buyers and Lebanese and
Armenians in Lebanon, Jews and Christians and Moslems, some Syrians.
And they would buy them and put them in milk bottles and bring them to
Antwerp, [where they] were cheaper than what they were getting from De
Beers. Cheaper is putting it mildly—maybe a twentieth of the price! And
there was one character who would give a very low price on diamonds
which were not perfect crystals and say, 'Look, these are broken goods.'

"Eventually the diamond dealers could do better with these diamonds.
Why were they wasting their money with De Beers? There were a lot of
sight-holders, but goods were coming in—were hemorrhaging, really, into
Antwerp—all those out-sights, as they're called, goods not syndicate-
controlled. And there were such bargains to be bought that the people who
held sights were buying those businesses and sending back their box to the

bank. They put their De Beers boxes in the bank. They hocked their boxes and used the money to buy goods from these Lebanese in Antwerp. They took the financing from the bank and used the money to buy their diamonds on the market. Eventually the banks began to bulge with boxes. De Beers didn't like [it] at all. They knew that if this went on, people would simply abandon their relationship with the syndicate. It would affect the price of rough.

"So De Beers tried to strengthen the borders between Liberia and Sierra Leone—which didn't work. You couldn't close the border in the jungle, you know. De Beers's espionage didn't work. You couldn't stop them. The only way they could stop the traders was they sent buyers [to Africa and Antwerp] to compete with the Lebanese. So the buyers started getting better prices, and the goods started flowing to the buyers. So then the clever ones in Antwerp took the rough that they got from De Beers and they gave them to [middlemen] to sell back, and De Beers started buying their own goods back. Those are called passport stones. Until there was some guy, super, super smart, from Antwerp who actually purchased some stones that had been cleaved. Someone who's got experience with diamonds will see that somebody made an incision to cleave the stone. And he says, 'Hey, this is something crooked going on here; next thing, they'll be submitting polished diamonds!' So they stopped that. Eventually the whole thing stabilized."

The following story also illustrates how people tried to skirt control of the cartel: "Sight boxes are sometimes sold as a package, and there are quotations for 3 percent above list, 5 percent above list, and so on; these quotations are 'in the dark' [sight unseen]. Now some sell it in the dark, which the syndicate does not like because it is too speculative. It breeds secondary markets which it can't control. Syndicate doesn't like when people make visible profits. When [the market is] very hot, the producers that they buy from will get harder to do business with. Now if it gets sold as it's normally sold—open the box, sort out the diamonds, then sell it—no one knows how much profit is being made. It's invisible. OK, so this man would buy a few boxes [and sell them]. We called him the Prince of Darkness. For the smaller sizes there was a big loss because they were overpriced by De Beers, many people felt, so they sold for 15 to 20 percent loss. And other stones, 5 to 10 carats, they gave a certain profit, and for big stones a bigger profit, and if you sold the whole box, you made a couple percent profit. This kept him alive. Everything was in his head, some very complex deals. He was some sort of mathematical genius, I think. Amazing. In the end he went bankrupt."

The stories bubbled out of this trader. "When you're telling diamond stories, you find time passes," he said. "There's a story about one Israeli who used to be a policeman and then he became a diamond buyer and went to

Africa to buy. He used to buy huge amounts of rough for a certain big dealer, and one day he was carrying $15 million worth of diamonds, and he landed in Paris, and couldn't make a plane to Belgium, [so] where was he going to stay with $15 million? So he went to the police station and he started kibitzing with the policemen and spent the whole night talking police talk. And the next morning he took a plane to Belgium. Israelis are unique."

Smuggling in Africa is aided by its long beaches of rich sites, and political unrest destabilizes political control. A 1960 report on the industry in Sierra Leone noted that only about a third of the country's production passed through official channels, and between eight and nine million pounds of diamonds were illicitly exported in 1959 alone, forming "part of the official export statistics of a neighbouring country, Liberia, whence they are routed towards the European markets" (Moyar 1960:29).

After World War II worldwide competition sharpened as Israel's and India's industries started to succeed. Several factors converged. Illicit diamond trading flourished because unstable post-war European currencies created conditions for black-market trade, and the United States imposed import duties and excise taxes on diamond retail sales. Though U.S. sales in cars, homes, televisions, and furniture produced a post-war decline in the market for gemstones, U.S. demand for diamonds rebounded in the 1950s. Diamonds went from being 25 to 30 percent of jewelry sales to around 60 percent in the 1960s, helped by new advertising and the growth of large retail chains that sold diamonds at reduced prices to the middle class (Shor 1993:106).

The power of advertising was monumental. In the 1960s De Beers took its marketing campaign to Japan and the effect was almost instantaneous. Never having purchased diamonds prior to World War II, the Japanese market proved to be fertile ground. By the early 1970s Japan had become the third largest consumer of diamonds (Shor 1993:132), and 60 percent of Japanese women received diamond engagement rings in 1981, up from 5 percent in 1967 (Harlow 1998c:213).

The rosy diamond climate continued to be plagued by threats, however. More sophisticated artificially colored diamonds began cropping up in the market in the 1950s and were quickly condemned by the WFDB (Shor 1993:132). Attempts to synthesize diamonds came to fruition when General Electric announced success in 1954. In 1959, it received a patent for them followed soon thereafter by De Beers. GE sued De Beers; the suit was settled with De Beers paying $8 million (Frantz 1994a). To attempt to control the new technologies, the GIA standardized the way stones were graded and assessed in the 1960s.

Increasing markets and new finds made a synergistic mix, and the CSO was stingy with distribution to keep prices high while the new producers were coming on line. Diamond dealers bought stones from the new African nations to supplement their CSO supplies. Meanwhile, new high-grade diamonds were found in Australia in 1979, and large diamond mines transformed Botswana from desperately poor to one of the fastest-growing economies in Africa. Despite the new sources of diamonds, the weak dollar in the 1970s, no longer tied to the gold standard, spurred rising commodity prices of diamonds, gold, art, silver, and platinum.

The diamond plenty demystified the gem somewhat for both consumers and traders. Against industry officials' wishes New York diamond dealer Martin Rapaport published diamond prices for all to see. Amid rumors of diamond shortages, the cartel increased diamond prices, and diamond investment schemes proliferated. When Israel stockpiled diamonds in the late 1970s, the CSO tried unsuccessfully to stem the rise in prices by putting double-digit surcharges on its sight goods. But as banks extended more credit, particularly in Israel, diamonds kept going up.

Then the United States raised interest rates steeply, to around 20 percent, and the price of a one-carat D flawless diamond suddenly hurtled from $60,000 to $10,000, sending the industry into disarray.

The crash was devastating to many. It coincided with the 1980s recession that decreased consumer demand for diamonds and other luxury items. Popular writers on the trade predicted the demise of the industry (Epstein 1981), and trade officials waged internal battle against Rapaport's price lists that reported the fall starkly. Zaire left the cartel. Public perception linked the diamond industry with the apartheid regime in South Africa, even though only a small percentage of the world's gems came from there.

After the 1980s retrenchment, however, the diamond market began to expand again. Fancy colored diamonds became popular and pricier. Larger diamonds were avidly sought while smaller diamonds found their niche in jewelry innovations such as Chris Evert's tennis bracelet rimmed with tiny stones. The auction of the duchess of Windsor's jewelry by Sotheby's in 1987 fueled new interest in jewelry and set off higher prices at subsequent auctions. The Asian markets for diamonds continued to expand, as did India's manufacture of small rough.

Breathtaking events in the late 1980s and 1990s also affected the diamond industry, and, as before, each development produced repercussions throughout the world. Relationships between the CSO and producers were key, and negotiations produced on-again, off-again deals. Angola stopped doing business with the CSO. Australia, the new big force in diamond production, had several contracts with the CSO, then decided to go it alone, prompting fears that manufacturing in India would be hurt if De Beers cut its prices to pressure Australia. The Soviet Union worked out its first agreement with the CSO in thirty years, then dissolved into fifteen republics in late 1991. Russia threatened to flood the market as it continued to negotiate for larger percentages of its diamonds to stay on the open market, but then, desperate for cash, reached agreement with the CSO. A report suggested that De Beers exaggerated the extent of Angolan smuggling to cajole Russia to stay committed to the cartel (Keller 1992). In 1990, Russia and Israel agreed to their own joint venture.

The recession, the Gulf War, and the U.S. luxury tax hurt business in the United States in the early 1990s. Many New York traders went out of business or left (Brooks 1992). De Beers slashed its dividends, its stock plunged, and negative publicity resurfaced in the form of television exposés about deceptive jewelry practices of fracture-filling and irradiation. *Time, Fortune,* and other magazines predicted the end of the cartel with headlines like "Diamonds Aren't Forever" (MacLeod 1992), "Hard Times for Diamonds" (Teitelbaum 1991), and "Falling Like a Stone" (Whitley 1992). Japan had a slowdown in its jewelry buying, adversely impacting world manufacturing centers further (see also *Diamonds: A Cartel and Its Future* 1992).

New diamond finds, along with expanding manufacturing and marketing centers, affected relations with the CSO. For example, new finds in Australia and Canada looked extremely promising, and Canada was expected to operate independently. Diamond cutting boomed in Sri Lanka, and Hong Kong, Singapore, and Bangkok became important diamond centers. India surpassed both Belgium and Israel to become the world's largest diamond manufacturer by weight while its jewelry business became the biggest foreign-exchange earner in the country (McDonald 1993).

The collapse of the East Asian markets in the late 1990s influenced the CSO to return to its policy of regulating rough "to insure the viability of the diamond markets over the long term by carefully matching rough supplies with polished demand" (Rapaport 1999). This policy, the agreement with Russia, and renewed warfare in Africa resulted in shorter supplies of diamonds all around. By the late 1990s, these developments led some to believe that the end of single channel marketing of the CSO might ensue in the near future. Russia was selling a large proportion of its goods on the open market, and Australia and Canada began to operate independently.

The CSO sales restrictions had the effect of boosting morale in the manufacturing centers of the world because firmer rough prices signaled the beginning of a diamond recovery on the supply side. Reports said that Japan was "coming back to diamonds," with early 1999 imports up 25 percent in dollars from a year prior ("Japan Inching Back to Diamonds," 4/21/99). U.S. demand looked strong as expansion continued, and though East Asian markets remained shaky, finances there seemed to rebound. To celebrate the new millennium, De Beers planned a $60 million advertising campaign (Rapaport 1999, Wagner 1999).

In the 1990s smuggling also continued to be brisk. Recent smuggling trade figures range from $200 million (MacLeod 1992) to $350 million (Melcher 1992) to $650 million (Shor 1996a:42). UNITA, the rebel movement in Angola, generated $3.7 billion from smuggling in the 1990s, according to Global Witness (2000).

Methods to sneak diamonds out of protected areas can be incredibly ingenious. One report described how a miner could press a diamond under his fingernail, then into his mouth, to get the diamond out of the area. Gas tanks, ears, and cuts in tires have been classic hiding places for diamonds, and homing pigeons and other means have been used as ways out of the area. Because x-ray scanners are used randomly to minimize the health hazard, miners can sometimes slip a diamond past detection. Once "a thief smuggled in the pieces of a crossbow, later sending a volley of hollow bolts freighted with diamonds arcing over the fence, for retrieval by a confederate. This scheme ended when an unlucky shot fell to earth in front of a security jeep" (Hart 1999). And though a place may be known for illegally obtained diamonds, proving such illegality has been difficult.

Matthew Hart, a mining correspondent, continues his account of sample security breaches connected with diamond mining. Beside a long stretch of the Atlantic from South Africa to Namibia, which is dubbed the Diamond Coast, lies Port Nolloth, a town that attracts illicit diamond buyers attempting to evade the official controls of the large diamond concern, Namdeb. "Consider the last raid," Hart writes, "which targeted the town's Portuguese, thought to be particularly active in illicit diamonds. . . . Vehicles full of heavily armed police officers sped north. A helicopter clattered into the air. They all headed for the Portuguese country club . . . surrounded by a wall topped with razor wire. The helicopter racketed over the wall and hovered. Combat officers slid down ropes into the green oasis of the grounds and charged the clubhouse. 'They had food,' says Derrick Clampett, a seasoned diamond detective who participated in the raid. 'They had food. They had booze. They were ready for a party. The diamond scales were all set up, there were loupes—the whole business. But there weren't any diamonds. We were a day early. Later we raided a house and found $250,000 in cash.'" Here another detective broke in with the punch line: "But the possession of money is not illegal in South Africa." Hart adds a sardonic question to finish this tale. "How much rough goes missing at Namdeb? The figure may be as high as 30 percent. . . . Then thieves are skimming off some $160 million a year . . . not a bad sight box, and they don't even have to go to London" (Hart 1999).

Diamonds also have been a positive force in nation building. Botswana, Namibia, and South Africa have benefited greatly from their diamond wealth prudently channeled within each country (Nessman 2000).[9] Canada's exploitation of its new Ekati mine will strongly enhance its economy. When Nelson Mandela became the first democratically elected black president in South Africa in 1994, he and Harry Oppenheimer formed a partnership between big business and the government that continued with the election of President Thabo Mbeki.[10]

War and the human devastation it causes tell a completely different story. War plays an important role in the diamond fortunes of various African nations, continuing the tragic irony of huge wealth that undergirds war amid desperate poverty. As Global Witness said, the paradox of the diamond is one of "enormous mineral wealth and devastating civil conflict" (2000:1). A World Bank study of forty-seven civil wars between 1960 and 1999 found that the single biggest risk factor for the outbreak of war was a nation's economic dependence on commodities, whether it was coffee, drugs, or diamonds. "Diamonds are a guerrilla's best friend," said the author of the study, Paul Collier (Kahn 2000).

When Angola legalized diamond digging against the backdrop of civil war in the 1990s, wildcat prospecting proliferated, and approximately 30,000 *garimpeiros* were involved in smuggling diamonds to a flooded Antwerp (Contreras 1992). Those developments led one to joke, "If there wasn't a war on, I'd have a proper job" (The rocky road to survival 1994:45). As usual, De Beers rushed to buy up the diamonds. In 1996, a United Nations-led peace initiative to end Angola's civil war encouraged shared control of the nation's diamond fields between the Angolan government and UNITA though the peace talks sputtered. In 1998, the United Nations instituted an embargo against diamond sales from the Angolan rebels that proved ineffective, and news reports described diamonds finding their way to western Europe, the United States, and South Africa (Daley 1998, Fisher 1998). Various peace treaties did not hold.

The economy, general welfare, and infrastructure of some African countries have deteriorated as a result of these wars. Liberia has no electricity, running water, government schools, or hospitals under the regime of Charles Taylor, who keeps himself powerful and rich while funneling dia-

[9] Namibia, Congo, Angola, and Botswana are some of the biggest producers of alluvial, kimberlite, onshore beach mine, and oceanic diamonds in Africa. Namibia and Botswana share ownership with De Beers. Other African nations that produce minor amounts of diamonds include Lesotho, Swaziland, Zimbabwe, Tanzania, the Central African Republic, Ghana, Sierra Leone, Guinea, Liberia, and the Ivory Coast.

[10] Though he was an outspoken opponent of apartheid who financed schools and affordable housing for blacks, one of Ernest Oppenheimer's Anglo-affiliated companies supplied police with ammunition and tear gas after the 1976 Soweto uprising (Schmeisser 1989:42). De Beers apologized in 1997 for having failed to house workers with their families, to desegregate workplaces, and to advance black workers (McNeil 1999).

monds from Sierra Leone to Western diamond centers (Onishi 2000). Meanwhile, a decade of war in Sierra Leone during the 1990s resulted in tens of thousands of people killed or maimed. A politician there commented, "Diamonds might have been our blessing, but they have turned out to be our worst curse" (French 1995). A three-month United Nations Security Council ban on diamonds from Sierra Leone in 2000 did little to constrain the $700 million-a-year trade there, however, as little will on the part of the affected governments was evident (Onishi 2001).

Increasingly, international attention has focused on the role diamonds play in war and brutality. The Global Witness report, *Conflict Diamonds*, issued in mid-2000, called for an organized international effort to reduce the flow of conflict goods and to "enable a diamond to once again be 'a girl's best friend.'" It labeled the biggest problem areas as Angola, Sierra Leone, Liberia, and the Democratic Republic of Congo. When international experts identified Taylor of Liberia as the main player in the diamond smuggling from Sierra Leone, Ukraine (mentioned as a source of some of the arms shipped to West Africa) blocked the report from being introduced to the UN Security Council in late December 2000 (Crossette 2000b–c). Burkina Faso was also linked with Liberia in the exportation of smuggled diamonds from Sierra Leone (Harden 2000). Other nations that have no diamond fields of their own, such as Gambia, can be in the smuggling business, too (Crossette 2000c).

Richard C. Holbrooke, then U.S. ambassador to the United Nations, compared Taylor of Liberia to deposed Yugoslav President Slobodan Milosevic. "Taylor is Milosevic in Africa with diamonds," he said, "fueling the conflict in Sierra Leone for his own benefit" (quoted in Harden 2000).

In response to this attention, the diamond industry, human rights organizations, and world leaders began to take steps to control the illicit diamond trade. South Africa expressed interest in developing an international certification system to stem the civil war financing as well as the adverse publicity (Crossette 2000a). The Clinton Administration considered actions coordinated with the UN, and De Beers said it would support tighter controls (Bonner 1999:3). An international representation of diamond producers, sellers, and De Beers, calling itself the World Diamond Council, worked together to eliminate the trade in diamonds that funds African wars. This industry unity, together with the Kimberley Process, a consortium of governments, recommended more effective controls and helped pass the Clean Diamonds Act in 2001. Their efforts gained momentum and urgency following the September 11, 2001 attacks when intelligence reports indicated that the Al Qaeda terrorist network had profited from the conflict diamond trade in Africa.

The preeminence of the diamond cartel has become increasingly more uncertain in the new millennium. De Beers now controls a much smaller proportion of production than it used to (Global Witness estimates 70 percent, 2000:1; others suggest 65 percent, Gaouette 2001). In 1999, De Beers reported that it had had the worst sales since 1987, totaling $3.4 billion (McNeil 1999). In November 2000, a De Beers rival won the bid to buy 40 percent of Australia's largest mine, Argyle. In 2001, amid fears of a U.S. slowdown and the impact of dirty diamonds on the trade, De Beers announced that its major shareholders were buying the company for $17.6 billion. No longer a public company, De Beers would brand its own stones as assurance that the stones it sold in its own stores would not originate from conflict areas. This move also meant it would stop stockpiling diamonds to control prices and would compete with its own customers for the lucrative retail market. Sight-holders would be required to increase their marketing efforts in order to remain sight-holders. One dealer lamented the change for the consumer. "The consumer will not get more diamond. He will get less diamond for his money, but he will get a nice package" (Weber 2001: 11).

With global conflicts, economic fluctuations, and the quickened impact of international events on the status of individual nations, the demise of De Beers has periodically been proclaimed through these years. Nonetheless, De Beers seems to be thriving. As the *New York Times* put it: "Periodically, the death rattles of the cartel, which sets prices for most of the world's uncut diamonds, are said to be heard . . . synthetic diamonds . . . Roosevelt's anger . . . Russian finds . . . Israeli hoarding . . . Australian finds and Zaire's defection. Each time, the cartel outwitted fate. While OPEC has withered and the tin cartel collapsed, diamonds have glittered on" (McNeil 1999).

The long history of diamonds testifies that events in one part of the globe intimately affect the fortunes of those in other parts of the world. Sometimes the effect is immediate; at other times it is subtle or delayed. In all, the globalism of this business has been present from the beginning, and in no less measure continues to affect locals in profound and diverse ways. And that includes the traders on New York's 47th Street.

◆ ◆ ◆

[3]

Being There

The elevator at 11 West 47th Street speeds to the tenth floor and deposits us into a large foyer. The men pull their ID cards out of their pockets, stride purposefully past the glassed-in security office, and swipe themselves in through the turnstiles. I exit the elevator, look to the guards, receive a nod from one of them, and pass through the turnstile that lets me into the club. Moishe, or my other uncle, Shmiel, would meet me there, and I would go with one of them to mingle and watch the trading.

Eleven West 47th also fronts on gleaming 5th Avenue, and, besides containing hundreds of diamond offices, is home to the central marketplace of diamond dealing in the United States, the Diamond Dealers Club, or DDC. The DDC is one long, large room on the tenth floor. (The DDC offices are on the eleventh floor.) This is the marketplace in the city where members meet to show, buy, and sell diamonds. The room itself is bright and airy. Tall windows form the north and south walls. Long tables form perpendicular prongs from the walls. Each table has four sets of fluorescent lamps, two at each end, and a half dozen or so chairs on each side. The tables are gray; the carpeting is gray; the uniforms of the staff are gray. Everything is visually muted, the better to examine the merchandise. A wide corridor separates the two sides of the club. On each side of the corridor is a shoulder-high wall lined with benches and telephones facing the tables. One proceeds down the corridor toward the glassed office where the diamonds that have been bought and sold are officially weighed and entered.

Above the glass walls of this office are clocks showing the time in New York; Ramat Gan, Israel; Mumbai, India; Johannesburg; Antwerp; and

11 West 47th Street / 580 Fifth Avenue is home to the Diamond Dealers Club of New York. (Photo by author)

Tokyo, some of the cities where diamond clubs (or bourses, as they are officially known) are located. The corridor narrows and then bifurcates at this point. Along the right passageway one passes bulletin boards detailing information about such matters as people applying for membership, out-of-town buyers, diamonds that have been lost and found, election information, members who have been suspended, arbitrations, and the latest scams. Also on display are notices about particular diamonds for sale or sought by out-of-town buyers, announcements of DDC meetings and committee meetings, special hotel deals for itinerant traders in various cities, and security information.

The other corridor leads to the cloakroom, the bathrooms, and the glassed-in office where the operators sit who relay phone calls to members. Behind these rooms is the room of lockers and the little *shul* where the religious members assemble to pray each afternoon. Both corridors empty into a wider area at the back where tables are provided for playing cards and reading newspapers. Behind these tables is the smoking room. At the right of the small corridors is the dining room and, angling farther to the right, the luncheonette, both businesses staffed by Israelis.

On this day some individuals are ensconced in their usual seats, most often directly against one of the windows. For these few, the club is their office. They set up scale, tweezers, loupe. Their goods are nearby, in thick wallets with numerous dividers for individual packets of diamonds, or in briefcases near or on the table. They avoid the overhead of a separate office, and they are in the thick of the New York market here. One man's business, for example, is to evaluate whether stones should be recut or otherwise refashioned before being sold or officially evaluated by the GIA.

Early in the morning there is little happening in the club. The few who are around are checking their inventory, readying their goods, sorting memoranda, or awaiting customers while reading the paper. As the morning wears on, more people track in and out. In addition to the people who walk the corridors, bunches of men and a few women at the tables talk, scrutinize goods, trade. The loudspeaker calls members to answer phones. As more people enter the club, the atmosphere begins to charge with buzzing and activity.

The two or three hours around noon are the busiest time of the day. The club allows its members to bring guests to the dining hall, and these people lend a festive mood to the room. The women are dressed up; their colors enliven the place. Midday religious services are full during this time, as well. Chess and card games at the back of the club, played mostly by older members, are boisterous and busy.

The hall jangles with the clusters of men, some women interspersed, intent, nervous, showing and examining diamonds, watching, asking questions. Traders hang on the club phones, slam receivers down hastily, confer with others. The business is hard and getting harder; the traders are edgy, competitive, trying to find a stone for a buyer, get a stone placed for a seller, trying to get a commission, trying to make a living. A trader gets back on the phone. Loudspeakers call names; men and women jump up to answer their pages. Increasingly, cell phones ring their distinctive chimes and more people walk purposely with phones pressed against their heads, speaking loudly, gesticulating, and clapping them shut, then opening and quickly tapping in new numbers for a fresh round.

On one of the first days of fieldwork in 1986 as I accompanied Moishe and Shmiel into the club, I was struck by the amazing array of long and stunningly white beards that dramatically contrasted with the heavy black caftans that the Hasidim wear. The club looked to me that day like a huge, boisterous gathering of extremely old men. As I continued to look here and there, I could discern within the sea of gray-white beards blending with the gray walls and carpeting men with black beards and men clean-shaven—in fact, men who were middle-aged, some even young-looking, and even a few women!

As if seeing my thoughts register across my face, Shmiel commented that people don't retire in this business, because they are able to modify their hours. Men can come in at 12 and leave at 5. In that respect, Moishe added, it's a relaxed business. And when people actually do retire, they still like to come to the club to play games, to socialize, and just to be around. Under discussion at the time were membership dues for those older than 80; a few years later it was determined that members over 80 should pay 80 percent and those over 85 would be free.

The club stays active until mid or late afternoon, then the bustle tapers off and quiets. Each individual has a routine, and the week has its pace: Monday is relatively slow, Tuesday through Thursday are busy times, and Friday is a dress-down day; the day is shorter and less active because the Sabbath begins at sundown. There is a seasonal rhythm to the year as well. Auctions and jewelry fairs in various large cities punctuate the year and determine the flow of the club as goods are given out and numerous traders travel to each site. Summers are typically slower times; during the fall the pace of trade picks up as the Christmas season beckons. Between January and the spring commemoration of Passover, traders take stock of the past holiday season and stabilize a routine. Many traders take vacations during the Passover month. That the club closes for *Shabbes* and every Jewish holiday is a recent concession to the increasing numbers of Hasidim in the business. Though it is hard to believe now, about forty years ago the club used to be open on Saturdays.

The DDC was the focal point of my research among some of the people who work in the diamond industry in New York City, particularly dealers and brokers. It is the central point of interest for many of them as well.

Because the 47th Street district is too large to study comprehensively, I mostly confined myself to the people who use the DDC as a pivotal center of their activities. I struggled with how to circumscribe the subject to make it manageable to study while acknowledging connections to the outside, to other times, and to the ethnographer. An essential arbitrariness helps create a sense of coherence, but to a large degree it is merely a pragmatic truce with only a part of reality.[1]

For many people, the club was their main marketplace, social center, hub of activities, and pulse of their business. Some people I talked with, however, rarely if ever used the club, but I include them because they are part of the network of traders. Besides traveling throughout the United States, Europe, and Asia for goods and customers, they rely on faxes, e-mail, and the Internet in their commerce. Some of the people with whom I discussed

---

[1] Anthropologists worry about boundaries of their studies and what the "field" means (See Knauft 1996). For Kugelmass (1986, 1988), senior centers and other gathering places in cities provide convenient ways to situate a study, especially in hard-to-bound cities. The "field" misleads by connoting a pastoral setting, as Gupta and Ferguson (1997:8) remind us. In studying the homeless in New York, Passaro (1997) resisted focusing on just one particular "site," such as a homeless shelter. Like the homeless, diamond traders often travel widely.

the business are based elsewhere; the New York DDC is just one of the places they go, and I caught them on some of their jaunts here.

Though most have offices outside the club, traders come here to touch base, get a feel for the market, socialize, be seen, and exchange news. I would accompany my uncles and my cousins here as they encountered various friends and colleagues through the day. I would stand with one of them in the corridor and listen as they exchanged news, argued politics, questioned me, and—incidentally, it sometimes seemed to me—showed one another diamonds. My uncles had vouched for me, so I was sheathed in the protective wrap of their reputation.

◆

I started fieldwork in 1986 by phoning Shmiel and Moishe and asking their permission to study the business through them. They were each puzzled and dubious. It is no different from any other business, they insisted—a refrain I often heard from others, particularly at the outset of the research. But my uncles agreed to tolerate my presence and offered to put me up for the few days that I would be in New York each time I arrived for a stint.

I had never been particularly interested in diamonds or expensive jewelry. Being a child of the '60s, I eschewed ostentatious jewelry, preferring ethnic beads and earrings. Unlike other relatives in the family, my husband and I didn't trot over to one of their offices to look for a diamond for a ring when we became engaged. (Paul had inherited a stone in an heirloom ring from his grandmother.) But when I presented myself to my relatives to begin this study, they each took a look at the stone in my ring, pronounced it an old-miner cut, a cut phased out years ago, and dropped my hand unceremoniously, no comment.

My access to diamond traders and manufacturers was made possible by the good name that my uncles had created over the years. Moishe's niece, pondered one of the men I introduced myself to. That carries a lot of weight, he told me. Your uncle Shmiel . . . a real genius, especially in rough. He can take a piece of junk and make something out of it. Transform it.

I was a little cowed and awestruck by the nobility this family seemed to represent. I knew they had a good reputation, but I hadn't known how good. My connection with Moishe and Shmiel enabled an ease and candor among these informants. Some must have been assuaged by certain assurances Moishe gave them. For example, I promised not to use anyone's name. Furthermore, Moishe would read my manuscript so that he could correct my mistakes and veto and/or modify the inclusion of sensitive subjects, namely security concerns in this sometimes extremely dangerous business. There were jittery jokes about the unnamed "secrets" that I might disclose. Though I learned no secrets, I was subjected to the general wariness given outsiders since I was not part of the group.

This also meant I had the shadow of Moishe looking over my shoulder as I attempted to put the ethnography down on paper. "Get the tone right!" he yelled in my ear. "Don't mess it up like others always do!" His reminders echoed in my mind, causing me bouts of writing anxiety. I reassured myself that I had the right and the obligation to interpret this material. As it turned out, he approved my first draft and wrote minor corrections on his copy, but he died before we could sit down together to discuss it carefully. The other traders in the family also read, commented on it, and approved it.

I had no idea how typical my uncles were of the people in this business. They were part of the old-guard European group distinguished from the American-born, largely Hasidic contingent of diamond traders. I was a non-native native, wondering about the "otherness" of myself, my family, and this familiar yet thoroughly foreign world. I lucked into this fundamental location because of my relatives and therefore was regarded with serious interest. With every introduction the significant reference point was where exactly on the kinship map I stood with my uncle or my cousin. This allowed others to locate me, and I was validated and made reliable by this process. Once I was introduced to someone who knew my mother as a young girl in Antwerp before World War II. Someone else told me that he had just seen one of my mother's cousins in Antwerp. Some of my relatives' good reputation reflected on me, was a mantle I had to show myself worthy of. Enmeshed within family connections, I had entrée but I was nonetheless an outsider unused to the language, the customs, and the routines that were natural and assumed for them.

In some respects I was like the *goyische yid* that Belcove-Shalin described herself as when she, a non-Orthodox, single Jewish woman, attempted to do ethnography among the Hasidim of Boro Park (1988a). We were Jewish, just not Jewish enough. Because she was seen as convertible and therefore teachable, her informants talked to her and explained. Though no one pushed a religious identity on me, jokes were often made that I was learning the business. This was a well-worn path, trod by authentic apprentices who strained to learn the ropes and the nuances. Though Belcove-Shalin was subject to matchmaking attempts by her well-meaning informants, people mischievously asked me when I would be joining the DDC. Surely a discount in membership could be arranged because of my special circumstances. Perhaps I was my relative's bodyguard for the day, someone offered. The joke that I was apprenticing was partly serious. Belcove-Shalin considered herself to be liminal because she was potentially Hasidic. I was liminal by being an outsider with some claims to insider status, not easily categorized.

Like ethnographers who have wondered about their native status and have criticized this kind of term for continuing artificial polarities, I reflected on the looking-glass puzzle of being the other at home gazing at the other within and across from me. Our identities are too complex to sustain these rigid divisions, however. Especially in this increasingly globalized world, we are more and more hybrid, hyphenated, or "ethnic halfies" (Abu-Lughod 1991). I agree with Narayan, who aptly writes, "Rather than try to sort out who is authentically a 'native' anthropologist and who is not, surely it is more rewarding to examine the ways in which each one of us is situated in relation to the people we study" (1993:678).

So much of what I saw, heard, and discussed was both familiar and strange. I felt a certain comfort alongside an ever-present unease of not fitting. My vague knowledge of the business consisted of a feeling of familiarity, a sense of "we-ness" about the people I was mixing with. But I knew less than the women in my family who grew up or married into the diamond business. When I visited relatives, I understood that their questions and comments presumed a knowledge of the diamond business that I did not possess. I was thus native in some immediate senses and extremely distant in others. The conversational style was familiar—New York Jewish, aggressive, fast, European-derived, Yiddish-inflected, Yiddish-mixed—in some ways more familiar than the bland brand of language I grew up with in Connecticut. Unlike Mitchell (1988), a self-styled *"goy* in the ghetto" who was embarrassed and overwhelmed by the Jewish linguistic and cultural style, I enjoyed hearing and participating in it.

However, I knew almost no Yiddish and was thus barred from all kinds of jokes and expressions. As they sometimes translated, I felt the sting and the ache of not knowing Yiddish. You don't know this expression? Ach! It's so good. Let me see how to translate. It means . . . Oh, it's really untranslatable. Nothing can really describe it. It means . . . but it's so much more than that! You have to understand the Yiddish. You sure you don't know Yiddish? A Yiddish expression that means "the one and only son" is used to explain how each diamond should be cut individually. A distant relative expressed her misgivings about my lack of Yiddish: She'd heard two traders use a Yiddish phrase at the club the other day that translates as "your hand gives power," meaning the man had good luck on his stone because it had come from the first person. I was concerned too.

Family members had varied reactions to my project. At a family function in one of my relatives' homes around the time I began, a few people approached me, smirky looks of amusement capturing their faces. Shmiel introduced me to the people I'd known all my life: "This is Renée, of 47th Street." A few wives told me that I would never learn anything since the business was too closed. The men in the family don't talk and the business is secretive, they crowed, so my project would be doomed from the start. The women seemed faintly charmed by my naiveté: I would presume on this male territory and find out the goodies that they were excluded from? And didn't I hark from my mother, the rebellious youngest child and only daughter, the one member of the family who had virtually fled New York to live in suburban Connecticut, an entirely other world, years before? Quaint that her offspring would find their activities to be of anthropological interest. Was I attempting a nostalgic foray, trying to come back to the fold that my mother had rejected? Defensively, I told them their reactions would become part of my study.

My uncles and cousins as well as others in the business were knowledgeable about anthropology to varying degrees, and they joked about the "jungle" that I had stumbled into. Many were well-read and well-traveled; they went to movies and art exhibits, and were thoughtful and questioning. I had to account for myself and my enterprise throughout the time I was there. Moishe would ask periodically, do I have an angle for my study yet? As research dragged on, wasn't I about through gathering data by now? Wasn't I beginning to hear the same information from each person? The critical shadow of my relatives' oversight was unnerving and bracing. As I attempted to answer their questions, knowing that I wasn't sure of my responses, I tried to use the questions to help define and develop my search.

Many in the family were in the business, either peripherally or centrally.[2] I learned that some of the women had had some interest in the business but they were early shunted away from it. We in my assimilated Connecticut family were never considered possibilities for recruitment into the business. Family members in Antwerp still work in diamonds. One New York cousin worked for one of the Antwerp men for a summer forty years ago. An Antwerp relative immigrated to Israel in the '60s, worked in a kibbutz, got married, and became a successful diamond merchant there who travels to New York periodically on business. My grandmother's sisters had husbands who were in the diamond business, and one of the great-aunts kept a diary about working in her father's shop.

---

[2] In addition to two uncles and five cousins who had offices in the district, a husband of a cousin's cousin worked in one of the exchanges. So did the cousin's aunt and uncle, and the cousin's cousin helped out. A few female cousins had tried their hand at jewelry but it didn't work out. A cousin of my mother's, long since deceased, had been a dealer and had ground diamond powder, which he sold to a diamond polisher. Before he died, he gave his business to another of my mother's cousins, who had been a watchmaker. He ground the diamond powder until a few years before his death in the late 1990s. Two of my aunts had worked for a time in their husbands' office, doing various office jobs. One of the aunts had worked in another diamond office for a while, as well. Some of the husbands of women in the family had tried the business for a year to several years, but all of them found other work outside the industry eventually. Various of my cousins and the male children of my cousins have worked in one office or another during summers to test whether they had any enduring interest in going into the business.

In 1986, I traveled to New York each month from my home in Massachusetts and spent two to four days in the city with my relatives there, staying nights at their homes. I would arrive at their office, lugging a backpack (and later a laptop) and dressed for the club though still wearing running shoes that I'd needed for the trek from Penn Station. The dress code at the club required that women wear skirts, dresses, or suits—businesslike, dark, prim, and severe. Pants of any kind were out. I would plunk down backpack, change into flats or low heels, and stash my notebook and pocket-sized tape recorder (always hopeful someone would let me tape) into my pocketbook, ready for action.

If I arrived between 8 and 9, the office was just getting started. Initially two uncles and two cousins would greet me; after Shmiel died, the greeters were one uncle, two cousins, and the son of one of the cousins. They would ask quickly about my family, and I would reciprocate. Important news shared, we'd lapse into silence. I sat on a chair in a corner, waiting, and watching. I felt uncomfortable and in the way; sometimes I fingered through a trade magazine discarded in a stack nearby, filled with shiny ads of lustrous diamonds.

For the next half hour or so, I'd listen to them prepare for the day. They retrieved messages on their telephone answering machine and asked one another if follow-up were needed. They went over inventory, made phone calls, reviewed yesterday's activities. What happened with the 4–96 pear shape? He's waiting on it. Well, Henry might be interested. He'll tell me this morning. Grunted acknowledgment, back to the books. When is David going to pay? Do you have the 8–11? No, I gave it to Chaim. Oh, I was going to show it to Stu. Let me have the three round stones. I'll take them to Gregor later on.

A telephone message makes them laugh. "Hullo. It's Moishe. I have something for you." Click. Moishe? Which Moishe? In these days, when the number of Hasidim in the diamond business is increasing rapidly, Moishe, Shmuel, Yankel, and Ari are the common Yiddish names. The traders must know a dozen Moishes. A last name on this recorded message would have been appreciated. They'll find out later, probably at the club.

The high-ceilinged office, two rooms with vestibule and "mantrap," is laid out simply. The main room has two long desks, jutting perpendicularly from the outside wall with windows high above. Two men face each other at each of the desks. On the wall where I sit is a shelf, which has a fax and copier machine, our paper-bag lunches for later, and various books and trade magazines. A couple of old Flemish prints and an oil painting adorn the walls, as well as a few other pieces of artwork that depict aspects of the ancient diamond trade. The back room contains another, mostly unused, desk with word processor on top, a sink, a water dispenser, some shelves, some cabinets.

The mantrap, standard for diamond trade offices, consists of one closet-sized room with a door at each end. To enter from the exterior, you ring a bell, and wait while someone on the inside sees you via a window and mirror, and buzzes you into the mantrap. When the door closes behind you, a second buzz, activated from the interior of the office, lets you into the office.

A third room, now unused since around 1990, was the factory shop that occupied about four workers. Diamond polishing machines would whir here. The foreman would take a stone to my uncle periodically during the polishing process for a conference on how it was proceeding, and then he'd retreat back to the shop to put the stone onto the wheel for more polishing.

While similar to some of the offices I visited, this one is different from many, as well. Although this office is simpler and smaller than most, others are even sparer and tinier. Some family firms consist of more members, and the offices contain more rooms, more people, secretaries, a lot of bustle. A number of the offices we visited were lavish. They included elegant furniture, elaborate prints, plush carpeting, artwork from Israel, various Jewish paraphernalia, special computerized weighing machines, microscopes, and computers. Flashy posters advertising diamonds were often displayed: sumptuous diamonds draped languorously over a ripe pear, for example. Posters like this, so slick and glamorous, shocked me with the contrast from the gruff approach the men took to the stones, goods like any other trading merchandise.

Some of the offices whose members were Hasidic had large posters of Hasidic leader Rebbe Schneerson, as well as other religious and decorative items. Iranian, Indian, and Israeli-occupied offices were often adorned in Jewish artwork from those countries. A contrasting office, the only non-Jewish office I visited, was vividly different mainly for the hunting memorabilia that were everywhere: pictures of moose and deer, a hunting cap, dark wood paneling. I wrote in my field notes of that visit:

> Hal Fiorentino opens the door for us, looks a little alarmed at me, but lets us both in. The office has stuffed game heads all over the walls, ram and deer and big fish. By his desk are taped pictures of the Virgin. There are photos of fishing and hunting expeditions and an NRA plaque. Not too like the paraphernalia I've gotten used to. We're here to return the 2–33 my cousin got on memo— in other words, loaned on consignment—from this dealer, who says he's been in the business for over thirty-five years and is getting sick of it. As they pass a series of stones back and forth across the desk, they compare vacation plans.

In my family's office I would sit and wait for the morning routine to be done. Then, before they quickly scattered to their individual rounds, I would muster my nerve and pick an uncle or cousin to ask to accompany that morning. Often I felt like a pesky intruder, partly a holdover feeling from being the younger cousin who wanted to shadow the bigger kids. But one or the other would indulge me, and off I'd go with him. As I knew more people on my own, I used the office time to set up appointments for interviews.

I also used the morning time to ask questions. I'd stop them while they were discussing something to ask them to explain what they were laughing about, disgusted about, or heatedly debating. My early notes relate an exchange in the morning:

> They are talking about a guy whom he won't sell to because he won't come up enough. They see my quizzical look. They explain: they're insisting on 6–7 percent below list price. The other guy called back with the same low offer and finally, after repeated refusals, the firm said: no more calls. One predicts the buyer will be back, but the others aren't so sure. They decide they have to stay high. One goes over the figures on his desk calculator. This is our cost, he repeats. We have to make a profit. Renée's writing this down, one says. The others shrug—so what?

When I returned from conducting an interview, they often wanted to know what I'd found out and what my impressions of the person were. They would offer some history, colorful background about the person, a new context to differently situate what I'd just heard. For example, they might say: Yes, we used to deal with him, but he still owes us quite a lot of money that we will probably never see. He had quite a hard time of it, we understand.

The office was home base. For me it was home base in the sense that when we'd rejoin, we'd compare notes, I'd ask questions, and we'd continue from there. For them it was home base because that was their hub of operations, where the answering machine was located, where they kept their books and some of their goods.

The DDC was also a familiar place, a central hive. Men and women ranged between tables or stood in knotted clusters in the wide corridor, pulled here and there by a beckoning person, a new buyer, an idea where the suitable diamonds might be shown. They trod their usual pathways in the club, well-worn routes back and forth from table to corridor to another table to another table, sitting, standing, walking.

A final home base for me was at Moishe's in the evening. Though I stayed with a number of family members at various times, my most regular place was here. Often, Moishe would have left the office or the club earlier than I did, especially if I had an interview or appointment scheduled for later in the afternoon. I would try to get back to the apartment by 5:30 in order to join him and my aunt Gisele for a drink at the end of the day, an evening ritual that we relished.

Evenings with Moishe and Gis always took the same form. Entering the apartment, I would shake off coat and backpack and set down the laptop, kiss them hello, give them jam from our garden, then go to "my" room to quickly change into jeans, comfortable top, and socks. They'd pour a drink—they would have waited—and we'd sit on the sofa to talk. Each of them would have a different section of the *New York Times*, and they'd ask: So how did you do? What did you think of so-and-so? We'd rehash the day, and I'd sort through some of my impressions, jumbled and conflicting. I kept most back, not wanting my developing impressions to be inhibited by their strong opinions and out of respect to other informants and their privacy. I'd be full of questions. Moishe would take a deep breath and launch into his exegesis of the arcane diamond world.

Moishe would also quiz me, tell me that I was making too big a deal of some line of thinking. He'd provide a perspective that he thought would set me straight. I think you're overestimating the thing about women, he might tell me, and then say why. And I would think about it and ask questions, digesting this perspective. He always confronted and challenged me.

After dinner we'd watch a little of the news, then Moishe would put a string quartet into the CD player and we'd have a few hours of individual work, reading, or discussion which sometimes lasted late. We favored Schubert for months. We had peppermint tea around 9 P.M. A few times I interviewed Moishe on tape, and Gis joined in. In later years I'd clack my field notes into the laptop, fleshing out the scratched notes of the day into more fully realized sentences, trying to keep them as fresh as possible. Doing these notes would often provoke the questions I'd interrupt Moishe with. Gis would do her needlepoint or read. Moishe would read *Commentary*, the *New Yorker*, or the *Times*, or get back to one of his other reading projects at the time, a work by Isaiah Berlin, the *Iliad*, or St. Augustine's *Confessions*. We would each eat a piece of dark Swiss chocolate before bed.

In the morning Moishe and I would prepare our lunches, ready our tokens and change, kiss Gis, and head to the office, walking as much of the way as possible. My uncles loved to walk, and walking is an apt metaphor for the business.

Walks at the beginning and end of the day framed the research and business day. The trip to the city with Shmiel from Queens would encompass a half-hour walk to the subway and a subsequent half-hour ride to our stop. He was nearing 80 in 1986, and walking was a lifetime, loved activity. We would leave around 8, walk briskly through the streets of his and the neighboring town, remark on the stately trees, watch schoolchildren scurry with fretful parents following, calling. We usually stood during the subway ride since it was rush hour. The time we spent walking and riding the subway was a good time to discuss questions, elicit his perspective on various topics, and elaborate on other subjects. Our discussions were diamond talk interspersed with family talk. Then we would emerge from the subway, jostled and pushed by the rush-hour crowd, and maintain our pace with the horde. Up the stairs, through the turnstile, and around the corridor, and suddenly the rush of Manhattan was fresh against our faces, always startling, always somehow a shock. We were on 6th Avenue and 47th Street, suddenly amid the bright and dingy storefronts, the crowds of liquid-moving people, the vivid colors of their clothing, and the extended flyers pushed into us and dropped as we made our way into the narrow building and squeezed into the crowded elevator.

Moishe lived in Manhattan and would customarily walk the almost forty blocks to 47th Street and back again at the end of the day. In late 1986 he was almost 75, but as he entered his mid-80s he had to limit his walking to a quarter of that. We used to walk to 79th Street from the office on 47th and then either take the crosstown bus across the park or walk through the park to the upper West Side, depending on the weather and the approaching darkness. The reluctant compromise my uncle resorted to in recent years meant less walking altogether. Leaving 47th Street we would walk north on 5th Avenue, pausing to admire the latest floral display in Rockefeller Center or the magnificent diamond necklaces in Cartier's window. We'd turn east at some point and go north again on Madison, where we'd pick up a bus ten blocks or so later. We'd grab a transfer for the crosstown bus, then get off at the other side of the park, to walk the rest of the way to the apartment. Mornings had been the long walk through the park, emerging at Central Park South and 59th and then continuing the rest of the way along 5th to 47th Street. Moishe's weaker knees in later years dictated that we pick up the bus nearer the apartment, transfer buses in order to cross town, and limit the walking to the last few blocks on 5th Avenue.

Walking is also the pace of the business to a considerable degree. Even in their own office they're up and down as they consult with their partners, go to check out a stone or to the vault to retrieve goods. It's down the hall, down the elevator, across the street, down the block (or several), into the building, up the elevator, down the corridor, in and out of the office. Then it's a trip to the bank down the street, another few visits to a few more offices, then to the club, and back to the home office at lunch.

Even—or especially—when the pressure is intense, one maintains a determinedly unrushed look. Walking is a deliberate cover for the churning pressure. A young broker told me that he used to rush around like crazy until his family told him to stop it because it didn't look right: he looked desperate, and that's not how it's done. Irritability, a regular accompaniment to the unease and uncertainty surrounding this volatile enterprise, is mitigated by the rhythm of this deliberate pace. The business is smoothed by *schmooze*, familiar talk that trails soothingly through the deals, harmonized by the walking cadence. A message at the office expresses interest in coming over, but you don't rush. You need to collect a check from a trader who owes you, so you remind him as he passes you in the street—but you don't look anxious about it.

You intend to meet someone later in the day; if you do, you do; if not, maybe the next day. A person will get back to you, but if not right away, so be it. The pace is measured, as if to quiet the anxiety thrumming at a constant low level within. At the club people sit and stand around in groups, or they sit by themselves waiting, available to look, to offer advice, ready possibly to purchase, to "give out" goods, perhaps to sell. People check with one another about personal news and get interrupted by someone looking for a particular stone for a particular customer.

What did you do with that flat pear shape? asks one smooth-faced dealer in a modern tie to another dealer in flowing black caftan and hat and the whitest of beards. Which one? answers the Hasid. The one that he offered on, the 6–24. Didn't sell it. The guy wants to see a 7, it turns out. The dealer has a 6–87 downstairs that he can get for him later today. The Hasid would be interested in seeing that one. Nod, OK.

Another person interjects his request, and when answered, joins the knot of people, or pokes into another group elsewhere. A broker reports to the owner of a stone the responses the stone is getting to his price. This social soup is the context of the proceedings; the interjection of a new joke or an inquiry about a stone is continually added to the lively brew.

Despite the deliberate pace, jittery anxiety and the adrenaline of worry and anticipation simmer underneath. The wives told me how preoccupied their men were, how nervous the work made them, how they would stew about their decisions. Talk is pressured as though it's the valve for nerve overflow. Interruptions, interjections, quips of humor, talking over one another are constant. Eruptions of anxiety transform into sharp jokes. Survival by wits is on regular display. Phones ring, loudspeakers call, and people are pulled first to one person, then to another, then remember to call another: What's up with the stone? Bargaining and jibes are pointed. Traders look for tiny advantages, worry that their vulnerabilities show, and know that the thin margins of success hang in the balance. The trade can be grim. Walk, don't run.

Each new advance in electronic equipment speeds business and heightens the stakes. Cell phones allow instant access and accountability. One person deliberately bucked the trend to get a cell phone, and it turned into a good strategy—until, realizing he needed it, he finally got one. While he was holding out, his partner commented, "He needs one, but he thinks it's better not to have one because it makes them sweat more when they try to

reach him. It makes *them* more anxious. It wears them down. [They're] trying to get in touch with him because he's not offering enough on the stone.
He's not offering enough, and they need to call him *now* because the owner
says he needs the stone back. And they say, 'Well, we can't get in touch with
him. He must be in transit. He's not in the club right now. He's not in the office right now.' It makes them sweat."

Every day we would go to the DDC. The procedure for guests—my official capacity in early months—is formal and stringent. A member accompanies the guest to the security desk in the front of the club, where there are
introductions, signings-in, and attaching a guest tag to clothing for the duration of the stay. A guest is allowed to visit during the lunch hours of 11 A.M.
to 1 P.M. only. They let me stay longer.

We discovered the rule forbidding non-members from being at the trading tables when a security guard informed us of this breach, first once, and
then a few more times. My uncles and cousins seemed unaware of the rule,
and they thought it a silly precaution to apply to us. However, the security
guards were unyielding in their insistence that I leave the tables each time
they saw me sitting there. Later, in anticipation of an upcoming visit,
Moishe secured permission for me to sit at these tables unrestricted. He
spoke to someone at the top and procured a signed note each time I intended to be at the DDC. Sometimes the note was late in getting prepared,
which meant that I was blocked from the club in the meantime. To get the
note, Moishe explained what he understood of my research to the DDC
management. This intervention took care of another procedural obstacle:
prior to getting the higher-ups' approval of my research, I was not allowed
to go to the club more than once a week. After Moishe spoke to management, the number of days I could visit was unlimited. Removing these hur-
dles made the work possible. As I became known and the security guards
understood the arrangement, they would let me come and go without the
accompaniment of a family member.[3]

The assumption is that the guest is in the club for a short time, most likely
to have lunch. In addition to guests who came for lunch, I saw an occasional
group of GIA students who top off their courses with a guided tour of the
DDC. Some non-members are regulars like out-of-town buyers who sport
tags with BUYER in large letters on their lapels. They sometimes take up
residence at a certain table for a month at a time, then leave for a month or
two before returning. Or they may be sponsored by DDC members,
brought in to buy and sell particular goods for a few days, or a week or two.

---

[3] After Moishe died in 2000, the closed nature of the business became newly apparent to me.
On several occasions I was curtly blocked when I asked for information by phone. However, if I
reached someone who had known Moishe, the door magically opened.

Buyers have special privileges and limitations, and they sit at the tables to buy. Many DDC members travel to out-of-town bourses to do the same there.

Though most people I met had been in the business for many years and were middle-aged or elderly, I also talked to young men and women who had been in the industry only a few years and were still getting acclimated to its routines and necessities. I talked to DDC officials, board members, and other high-ranking personnel in the industry, as well as to those who rarely frequented the club and those who were not members. I encountered people of different nationalities, ethnicities, and religions. Jews, the most numerous, ranged from those who were unaffiliated, secular, and/or atheists, to those who were more religious, including those who were ultra-Orthodox. Traders from East Asia frequented the club, and I met some from China, Japan, Korea, Hong Kong, and elsewhere. Other traders were from Israel, India, Italy, South Africa, France, Belgium, England, and Germany. U.S. traders from outside New York included those from Texas, Florida, Nevada, and California.

I almost never knew whether the people I spoke with were successful traders. Moishe called the bulk of them "middle-class," the convenient term with which most Americans label themselves. The traders did not tell me what they made, and I could not ask. Diamond merchants rarely if ever talk with one another about their finances and level of success, though I knew that some were clearly struggling. One Jewish tradition involves bad-mouthing one's own fortunes, complaining how terrible business is. This convention of politeness acknowledges a wary fear of upsetting deities with displays of excess pride, in the *keine h'ora* [against the evil eye] tradition. If they do talk among themselves about finances, I was not privy to this talk.

I heard stories and news about bankruptcies and people having difficult financial times that sent shudders through the diamond community and yet were shrugged off as occupational hazards. Some of the notices on the DDC bulletin boards alerted members to disappearances. Traders who couldn't meet their debts or pay for their goods sometimes just "disappeared" into the City or overseas, unwilling to face their creditors, who were left seething and financially threatened in turn. People knew how precarious the business could be, and even if they themselves were making a go of it, their livelihood still depended directly on others. The web of trade involves countless numbers of people.

People were interested that I was interested in their work. To their grave nods, I explained that I wanted to provide an authentic account of what they did to counter distortions I read about the diamond business. The trade was often portrayed with embarrassing sentimentality—the quaint Orthodox Jews doing their ancient unique craft in frozen, traditional ways—or it was depicted with muckraking zeal about the machinations of the De Beers cartel, the marketing hype of the essentially plentiful diamond, and the crime associated with the trade. An entire murder mystery genre about the diamond business, with varied characters, plots, and international locales that excite and regale, is indeed wonderful, lurid, and spicy, adding luster to the intrigue of this business.

As an example of the sentimental stereotyping, this excerpt from a profile of a diamond merchant begins as follows: "Through a storefront window, diamond cutters and polishers can be glimpsed hunched over traditional cutting saws and polishing wheels. Through another window, two thickly bearded, yarmulke-wearing men can be seen shaking hands and opening a bottle of schnapps—perhaps toasting the consummation of a multimillion-dollar deal in an old-fashioned industry where a dealer's word is his bond, and contracts and lawyers are rarely used" (Ruby 1996:38). When Moishe came across this passage, he pointed it out to me as typical of the exaggerated sentiment and distortion that the industry is subject to.

In characteristic form he bullied me, finger solemnly wagging at me, emphasizing that when I got down to writing, no matter whatever else I did, I should "Get the tone right!" He quoted the article to several of his friends, steaming, challenging them: Have you *ever* in your sixty years in this business seen Hasidim through a storefront shaking hands and drinking schnapps? They agreed it was a ridiculous characterization.

Anthropologists have noted the new realities of globalization, urging attention to daily lives in specific locations and carrying out "ethnographies of the particular" (Abu-Lughod 1991:138). As we witness and participate in a world careening rapidly into tighter integration, the manifestations of this process are felt in unique ways. Ethnography for me was composed of narrative description of what the diamond traders did, alongside the stories of how and why they did it. These narratives are the content that helps convey the diamond traders to others who do not know them or their world. As "global facts take local form" (Appadurai 1996:18; Featherstone 1995:9 calls it "glocalized"), the stories of individual traders illuminate who they are, what they do, how they perceive aspects of their lives.

So, adopting a phenomenological stance with the daily particulars of these traders, I explained that I wanted to describe what actually occurs among people in the New York trade. How do they view the business? Why did they follow their fathers into the business? How has it changed? Have opportunities opened up for women and if so, in what ways? What do traders talk about and how is business conducted? Would they want their children to go into this business today?

Doing the ethnography I was trying to learn the basics. I tried to recognize impressions that did not fit, that jarred, that resisted a consistent pattern in order to be open to the flux and contradictions in the subject. My lack of insider knowledge was helpful in that everyone knew he or she could teach me, that my "gaffes and potential breeches" with informants mimicked a kind of apprenticeship relationship that would aid learning (Kugelmass 1988:16–17). Sometimes I felt invisible and this was good. Sometimes I sat silently and attentively with an uncle or cousin as people looked over stones and talked. If my uncle or cousin were called away to look at a stone at another table or to answer a telephone call, I stayed put as new people sat down around me, conducted their business, and held conversations next to me. I sat, didn't make eye contact. I thought if I drew attention to myself, they'd go away. I was treated as invisible and was ignored. I listened to conversations, heated arguments, requests for stones, hagglings over prices, debates about Hasidic sects, plans for outside buyers. At these times I felt I couldn't take out my notebook, but I yearned to scribble or to magically press a record button on an invisible tape machine.

I couldn't use the tape recorder in the club, and I couldn't record when I was in private offices with my relatives. I wish I could have captured the conversations, the pestering and bargaining, the injections of jokes and stories in all their cacophonous jaggedness, as indecipherable as it all might have turned out to be. Occasionally, fortunately, individuals allowed me to tape interviews in their offices.

It was harder to schedule separate time with these traders. They were busy!, and time was precious. I often scheduled an appointment, only to find out that the manufacturer or the dealer was unable to make it because something had come up, an outside buyer was there, a deal was imminent. Illness, back problems, and other ailments canceled some of the meetings I scheduled with older traders. This kind of occurrence could not be predicted. It was business first. This anthropological hazard of being at the mercy of informants' willingness to take time to talk was a given. The *schmooze* and apparent languor that the traders displayed in the DDC belied the urgent, bottom-line necessity to make a living and to get the trade done, not to talk with me. So when Moishe was able to snare someone to spend time with me, I snapped to and whisked notebook into service.

The fact that people talked to me was constrained by a general wariness that others have described as secretive. A general distrust—of authority, of government, of outsiders—has frequently been the case with Jews through the years. Such suspicion has been warranted. For centuries, Jews were subjected to cruel and capricious treatment at the hands of non-Jewish authorities. However, though Koskoff (1981) has written that diamond traders are wary of police and do not cooperate with them, I did not hear this. On the contrary—one diamond dealer told me he was grateful for a significant police presence on the street, and he wished there were more of them to make him feel more secure. Others agreed absolutely.[4]

Moishe cautioned me against writing notes in clear view of the traders for fear that they would be inhibited from talking with me. So I listened to people trading and then quickly went to a nearby bench, the bathroom, or the lunch counter to scribble notes as fast as I could. In the evenings I would transfer my notes to word processor. As time went on, I decided to display my notebook and pen openly. After securing permission to take notes with each person, I would jot down what people were saying as they talked. I decided it was important to carry the notebook prominently to mark that I was attending to the proceedings, framing these events, and trying to learn and to understand. I was not a guest, not merely my uncles' niece and my cousins' cousin, and not just enjoying the chatter in a passing, insignificant way. The notebook and pen also signaled my willingness to be questioned by them about what I was doing there.

People were not put off by the notebook, as it turned out. After Moishe or Shmiel introduced me to a friend or acquaintance with an explanation that I was an anthropologist, he'd ask that person to speak with me. As time went on, Moishe developed a spiel for this purpose: "This is my niece, Renée. She is an anthropologist and is doing a book about the diamond business. She won't use names. I want you to talk to her and tell her everything about what you do and help her out." Sometimes when the person would launch into a subject that Moishe considered elementary, such as, "The Jews used diamonds to escape to freedom . . . ," he would interject, almost scoffing: "She's way beyond that. Tell her something she doesn't already know," before leaving us to our interview.

Departing from office and DDC, I accompanied each uncle and cousin as he went on what I called "rounds" from office to office to club. I saw the routine of each of these men as they went about their day. I observed their interactions with their colleagues in their offices. My uncle might go to an office to return a stone that he had been considering and was now giving back. Now he might have a request for a particular kind of stone, one of a certain carat weight, cut, and color grade.

Inspecting the goods was ritualized. As the other trader looked through

[4] Security may be more of a concern in New York than in any other diamond center because the diamond trading and manufacturing offices in New York are spread over a large area, compared with other diamond centers. In Ramat Gan, Israel, for example, the bourse, individual offices, vaults, and banks are contained in a large complex that has state-of-the-art security measures. Going to individual offices, vaults, banks, and club in New York requires travel in and out of buildings and up and down streets. Decades ago New York diamond traders used to conduct business on the street, and it was not at all unusual for traders to exchange goods on the curb (Lubin 1982). Still today, New York diamond traders conduct snatches of business on the street though sternly discouraged by security officers, police, and DDC management.

his small packets of goods, he would toss one, then another parcel to my uncle to examine. My uncle would deftly unwrap the thin, bluish-white parcel paper by flicking his thumb underneath the fold to quickly open the package, scrutinize the stone with loupe, then rapidly close the parcel, and toss it back across the littered desk to then open and examine the next one. As the other trader handed little packets of single or multiple diamonds to my uncle and riffled through his thick wallet for other parcels that might be of interest, the two would keep up a flow of personal and political news as well as comment on the stones in question. Each packet examined was usually returned with scant interest and a terse comment: "This is badly cut," "Do you see the fluorescence?" "The shoulders are up," "Not a good color," or simply, "Not for me." If the parcel engendered more interest, it was followed with questions as to price, quality, or terms, followed by negotiation and perhaps a tentative deal. They would also inquire about other goods that were sought for other clients, ask about or make payments to one another, return stones, take stones on memo, and tear up memos on returned stones.

Traders often take stones on memorandum in order to show them to partners or customers and to give a decision later. This is lending goods on a consignment basis, an informal but usual transaction that is time-honored and trusted. The memo pad comes out, a memo is written, torn out, and each person retains a copy. One time I was sitting in the office of an Indian trader who examined another trader's stones. The exchange illustrates the memo system: The Indian peruses, then dismisses, each stone in turn. Then he gets some stones to show the other trader. Back and forth they pass the parcel papers, backs hunched, inspecting with loupes. He pushes them back, no. Don't like. Not a nice make. That's too dark. This one not enough life. The Indian trader says he has something else that is a certain size, and the visitor says he might be interested. The Indian says he has a customer who needs a particular K or L, marquise, around 5 carats. The visitor says he doesn't have something exactly like that but has a pear-shaped H, around 7–8 carats. Yes, bring it over. The Indian appraises another stone, remarks, "This is more interesting; it is a sleepy stone," and turns it over in his hand to scrutinize the shape and life from different angles. He takes one stone to keep for a while. The visitor writes a memo and gives the Indian a copy.

The memo way of doing business works, and no one sees a need to change it. Zvi hands Nuchum a stone. Nuchum says he has enough emerald cuts and hands it back. He receives another. He looks at the specifications of what's inside—the carat weight and cut of the stone—then opens the paper, takes out his loupe, slides the diamond across the seam of the parcel paper into his hand, wipes the diamond along the bottom length of his pinky, puts stone to loupe, and examines briefly. Wordlessly, he closes the parcel, tosses it to his partner. Zvi tells him it's 42 percent below list. It's too expensive, says Nuchum. After a second, Zvi tells him another percentage, and suddenly and silently Nuchum reaches for a bank envelope from a pile at the end of the table, takes the stone, places it in the envelope, writes on it, seals it. Done.

To the typical accompaniment of "She's with me," a cousin would usher me through the security of a new office, a new world of different people, décor, and scope of business. In we'd go through the first door of the office, unlocked from the hallway. Once through that door, my cousin would press the buzzer and simultaneously peer through the glass at the secretary or trader on the other side. We'd be buzzed into the mantrap, then buzzed again into the office proper.

My cousin would then introduce me and briefly describe my project, and I would take out my notebook, place it on desk. The businessmen joked about anthropology and the primitives they believed I was investigating. How exactly was I related, they'd inquire. "She's my father's sister's daughter" was greeted with "Oh, OK." Because some of these diamond dealers, manufacturers, and brokers were not members of the DDC, and some members rarely came to the club, my seeing them in their offices showed me the increasingly large amount of business conducted outside the club. Often, people consented to my returning to conduct private, open-ended interviews.

In addition to face-to-face interviews and informal observation, I also conducted several telephone interviews from my home in Massachusetts. I often met or heard of people with whom I wanted to speak more extensively but was prevented from doing so in person because I had to return to Massachusetts. Traders appreciated phone interviews because they took less time out of their business day.

Doing ethnography in one place while living in another provided challenges and opportunities. Family obligations necessarily interrupted the ethnography and created frustrations at times. For example, I was in New York prior to an important DDC election of officers and witnessed the campaign excitement, but I was not there during the election and had to rely on the telephone for results and reactions. I would hear of upcoming events, such as Market Week at the DDC, that I would have to miss. The telephone, and later, e-mail, helped fill in some of the detail, but being there would have been preferable.

Despite the frustrations that the distance and time span conferred, the necessary detachment enabled reflection. Each time I went to New York I was struck anew by the blend of familiarity and exoticism of the people with whom I was working. The fact that the project stretched to encompass well more than a decade's worth of time meant that I witnessed some of the most important but subtle changes in the business during that period. Perhaps the duration of time was therefore "a blessing in disguise," as Sanjek has described long-term research that creates a favorable "time perspective" (1998:10). Understanding the pull and tug of constancy and change during this period would have been impossible without the perspective of the number of years involved. Nonetheless, the interruptions my life imposed were significant and unavoidable.

When I began the project I was pregnant with my fourth child, and when she was born I stayed close to home. The older children were 11, 10, and 6, soon to be teenagers. Paul and I wanted to be present for talking, listening, and supervision, for participating in the daily problems and delights of children, the summer camps and jobs, the college visits and decisions, the wisdom teeth and illnesses. Everyday chauffeuring to lessons, friends' homes, dances, and yearbook meetings, appearances at track meets, piano lessons, recitals, Girl Scout banquets, and cross-country dinners were part of the package. During this period, I held consecutive part-time teaching and research jobs.

In 1995 I renewed monthly excursions to New York for a second fieldwork year. This fruitful period, however, included concerns about my parents that culminated in the sad decision a year later to place my mother in a nursing home near our home. I became the point person for her care, and my work was derailed for a couple of years. As this unsettling shift in our lives began to stabilize, I made a few more visits to New York to do more ethnography and finally decided it was time to "write up."

◆

The two brothers, Shmiel and Moishe, were the second and third children born to my grandparents whom we called Bonmama and Bonpapa. The oldest brother, my uncle Aron, was born in 1902. Soon after Shmiel was born in February 1907, my grandparents, who had been born in Galicia (now Poland and the Ukraine) in the last quarter of the nineteenth century, immigrated to Antwerp, a diamond cutting center. Shmiel liked to say that they moved because he, at four months of age, objected to Poland. They ran a general store in Galicia. Though he had never done this work before, my grandfather, like many Jews at the time, purchased a girdling machine to start a small business in his home, assisted by Bonmama. This small investment in the machine made it possible to set oneself up in business. The dozens of jobs connected with diamond manufacturing and selling were among those available to Jews. Moishe recalled in an interview:

"He started [the work] in Antwerp. He just watched somebody. I don't remember what he told me, whom he watched, but he watched somebody for a week and then he started on his own, and he took some work on contract. And my mother helped him. . . . [She] worked a treadle, like a sewing machine, and that made the girdling machine work. And he was girdling this, and this is how they worked for about a year or two. [They] saved up a little very small capital and with the small capital, they started manufacturing. That means he bought some rough diamonds, very small quantity of it, and he gave it out on contract to be polished, tried to sell it at a profit, and this is how he started."

Bonpapa had a small room upstairs in their row house where my mother remembers seeing him sit for hours in deep concentration, a Bunsen burner with a small flame burning in the background. At other times he went to the Antwerp diamond club to trade. Moishe recalled later:

"Before I started, my father made his office at home; he worked out of his home. He was manufacturing, he was giving out on contract to Belgian cutters who came over to the house, and then he would go for a few hours to the bourse in Antwerp and do his trading there. But actually he worked out of his house; he did his sorting; he had one room set aside for that. And it worked; it was an office. There was a desk, a rolltop desk I remember. And he worked there. I remember him sitting all hours, sometimes until 10 at night, and he would sort. He manufactured very small goods, which required a lot of detailed work."

Moishe was born in January 1912. With the outbreak of World War I in 1914, the family was expelled from Belgium to Poland via Vienna because the area of Poland from which they came was controlled by Germany, and the family, consisting of German nationals, was therefore considered an enemy of Belgium. In 1915 the family was allowed to go back to Belgium, then was ordered by the government in 1918 to go to the Netherlands—and then, like most other families in their situation, was finally invited back to Belgium after the war.

My mother, Anne, whom the family still calls by her Flemish diminutive, Anneke, was born in late 1921, the "surprise" of the family. Her brothers were almost 20, 15, and 10 when she was born, and as the baby and the only girl in this male-dominated universe, she recalls being both cherished and dismissed. She remembers that her questions about the business of diamonds were disregarded. Though she was not expected to work with them, maybe a husband or son would be recruited in the future.

In contrast, each of her brothers left school soon after becoming *bar mitzvah* around age 13 to work with my grandfather, the natural next step for them. Aron started as a cleaver at that age. He thereby learned one of the two primary methods of dividing a rough stone in half, the other method being sawing. Though such an approach is less typical today, many men who entered the diamond business traditionally started learning the business by mastering one or more manufacturing techniques, such as cleaving, and moving on to the more demanding and detailed work of polishing the facets.

Moishe and Shmiel watched their father and brother do the work in the office in the home. Shmiel came of age around 14 and joined them in the office. Bonpapa moved their operation into separate office space outside the home a few years later, and Moishe quit school to join them when he was about 17, having learned to cleave at age 16. When Moishe was 17 or 18, he started doing other functions on his own, such as buying.

Though Shmiel told me little of how he learned manufacturing, he also said that he probably went into the business "out of ignorance," that he knew nothing else, and that sons naturally went into their fathers' businesses in those days. "Just like you come down in the morning and have breakfast, that was how you went into your father's business," he told me one morning at breakfast before we headed off to the subway.

When I asked Moishe the same question, he replied, "We didn't examine it. You see, it was more the case that we had no other options. We didn't examine the validity of our view. We just knew that this was the way for us to go because we didn't have any other alternatives. So your question was actually not a question for us."

Opining that the plethora of choices today creates its own difficulties, Moishe described how he started out. When he was young, he watched his father and older brother Aron work at home where Bonpapa had a small workroom. Moishe was 3 when Aron started cleaving. Then Shmiel came of age and started to work in the office, quitting school at age 16. A few years later Moishe was working for his father too. Soon he was doing the selling while Bonpapa and Aron did the manufacturing.

Though my mother related how she had not been considered for the trade, I learned from an Israeli cousin that his mother had worked in her father's Antwerp business in 1927 when she was 17 and kept a diary of those few months. "I've been out of school for one and a half months," Bertha wrote to explain the purpose of the diary and continued, "I'm rather sorry about that. It was, altogether, pretty nice at school. But now it's finished. And now Papa has decided that I—instead of going to work as a secretary at an office, as most girls surely do—will learn to cut diamonds with [her older brother] Wolf. I have already started today! Now I'll stop because it is Shabbat and I'll surely get a scolding for sitting up here scribbling for so long. Now, for a beginning it's gone very well."

Her charming diary provides fragmentary details of working with her father and her brother and is interspersed with news of the evening classes she took, the lectures she attended, squabbles with her brother and sister, the weather, the new film *Metropolis* by Fritz Lang, and her 6-year-old cousin, Anneke, my mother.[5] On her first day she writes, "Today I've already

[5] For example, the diary includes this entry: "Aunt had gone *davenen* [praying, to temple] and Anneke can't stay home alone. She's a beautiful child. She's been going to school for a few weeks and

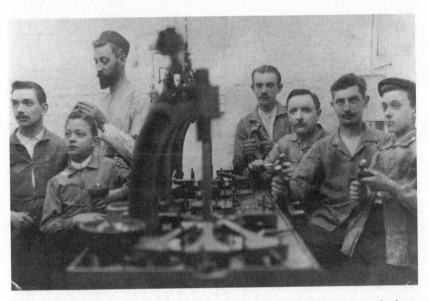

Bertha's big brother, Wolf (seated at left), with his father and employees at the business in Antwerp, circa 1910. (Author's collection)

Wolf and an employee take a coffee break in Antwerp, circa 1938. (Author's collection)

cleaved or cloven. How do you say that? Now, all beginnings are hard, as the proverb says, and as I've also experienced today. I hope it will get better soon and I comfort myself with the thought that I'm not the only one who has made it through the same thing."

By the end of the month, she sounded proud of herself. "I am now definitely a *kappeskliefster* [diamond cutter]. Nice title, eh? It goes very slowly, but I'll just think that it's 'slow and steady.'" She worked with her brother and two boys who were her brother's apprentices in a room in their home. Mostly, the work was tiring and confining, "sitting home all day." She contrasted this tedium with nostalgia for school. "All day I sit still: you obviously can't go wandering with the cutting bowl!" "The week goes by monotonously. Only the evening classes give it some variation. . . ." She sounded increasingly disenchanted. "The diamond-cutting is going so-so. Naturally, in the beginning you can't expect much." Her diary ended soon thereafter, and so did her diamond-cutting career.

In 1927, at age 20, Shmiel traveled to South Africa to scout the possibilities of a direct link for the family firm. Though Bonpapa was not in favor of Shmiel's going to Africa and was not able to finance the trip, Shmiel insisted on the expedition as a demonstration of coming-of-age independence. The diary recalls the departure. "There were some visitors, namely, my three cousins, Aron, Shmiel, and Moishe," she wrote, continuing in an enthusiastic and romantic vein.

"Shmiel is leaving on Tuesday for South Africa and came to say good-bye. A considerable trip! Just to Johannesburg. He says that'll be at least four weeks before he arrives. . . . It is really quite terrific to make such a long journey! To see something of the world, to come in contact with other people and customs, that must be so interesting! And then, after a few years, to come back, rich with memories and experiences, with more insight into human nature than can ever be learned from a book, while you know that there is still a house that remains open for you, a house that has never thought to abandon you, no matter how many things take up your time."

Bonpapa arranged with a more prosperous diamond dealer to provide the financial backing so Shmiel could buy diamonds in South Africa and ship them to Antwerp to cut and sell. This revolving credit arrangement was cut short by the stock-market crash of 1929. The world financial markets,

---

now her mouth isn't still for an instant. The whole time she was talking about school. Oh, and we laughed so much! She told me that Monday morning she wasn't at school for one reason or another. I don't know exactly why. And in the afternoon they gave her a note that she had a headache, which wasn't true! And Anneke said that the teacher asked her: 'Well, Anneke, did you have a headache?' and she answered: 'No, teacher.' 'No?' wondered the teacher. 'But doesn't it say that right here in the letter?' But she said, 'That's really not true and I'm not allowed to tell any lies, right?' 'But why did you stay home then?' the surprised teacher asked. 'Oh, I don't know. It was a party, I think.' "

especially the diamond market, fell precipitously, all things financial seemed to come to a standstill, hundreds of thousands of people were propelled into poverty, and the bourgeois world was rocked on its foundation.

Shmiel returned immediately to Antwerp and discovered soon thereafter that he had tuberculosis. My mother, only 7 when he returned, was shocked by his great height, his sunburned face, his loud voice and laugh, and his intrepid command of English. He seemed a stranger. My mother remembers the family's intense fear of the disease coupled with a palpable sense of shame and secrecy that attached to the illness. Shmiel recovered after spending about two years in the sanitarium in Davos, Switzerland, the locale and facility made famous by Thomas Mann's *The Magic Mountain*.

The family was in turmoil from these immense world and personal events. In the meantime, Moishe had started working with his father, and Aron had gotten married; he and his wife, Lana, lived next door, and the first of their three sons had been born. Shmiel and his fiancée, Rachel, got married in 1936; Moishe and Gisele married two years later.

Though they had intended to stay in Europe—Moishe and Gis had just bought the furniture for their new apartment—Moishe "had the feeling of some impending doom" and with the encouragement and insistence of his two older brothers, he and Gis decided to leave. Since my mother and Moishe had been born in Western Europe, the United States considered them quota-allowable. That is, the other members of the family, having been born in Eastern Europe, had to wait a long time to obtain a U.S. visa. Once in New York with the clarifying sense of objectivity that the distance and a different country seemed to confer on him, Moishe sent repeated cables, irritating and frightening the family in Europe. The cables pleaded for the family to make plans to leave because of the darkening situation for Jews that he could see happening in Europe.

Bonpapa listened. He told his sons that they should go; and he and Bonmama would stay. Aron told Bonpapa to make the decision for the whole family: if he and Bonmama stayed, the rest of them would too. Bonpapa then decided they would all leave. The family believes that Bonmama and Bonpapa would have perished in Europe had Aron not held firm and made sure Bonpapa was the decision-maker.[6]

Bonpapa gathered the necessary monies for fees and for the greasing of palms that enabled my mother's boat passage to the States in 1939 just months before Hitler invaded Belgium. Since she was only 17, she was sent

[6] This story contrasts with another: A father prepared his family to leave Belgium and wanted his parents to come with them. The parents said they were too old and refused to leave. The young family then immigrated to New York, and his parents perished in a concentration camp. The young man, now in his 90s, has always felt partly responsible for his parents' deaths.

to live with Moishe and Gis in New York. Moishe started a business in New York, working first in his father-in-law's diamond office (Gis's parents had left Antwerp a few months earlier), and then opening his own shop. For the first few years, business was very difficult, but a few years later the price of rough goods was low, and business improved.

Meanwhile, the other immediate members of the family secured passage to Brazil. As Moishe put it later: "I wrote to Aron to ask them to send the three boys. [Actually, because they were younger than 16, Aron's sons were not Belgian citizens, which made them ineligible for admittance to the United States.] But then they started looking around for alternatives they had, Cuba and other places. Then Brazil came up. It was easier because with money—a few thousand for a visa—if you could prove you were a capitalist and that you wouldn't become a burden—actually like here, where you have to give an affidavit meaning that you could support yourself. There you had to show money that you were a capitalist."

The family didn't have the money, but they were able to borrow it. Aron and Shmiel decided to immigrate to Brazil since it had a small diamond industry, and the family could continue its livelihood there. From there they were able to send goods to Moishe in New York, where he was establishing his business. Aron and Shmiel and their families then went to New York in 1941, and their parents arrived in 1942.

The Depression still lay heavy on the city. The bread lines were long and sinuous. The Diamond Dealers Club in New York, founded in the downtown area of the Bowery on Nassau Street in 1931, moved in the mid-'40s to 36 West 47th Street to accommodate the Jewish diamond dealers who had fled the Nazis.

At first Moishe shared an office in the current diamond district, though the Club was downtown. In 1940 he "put up a small shop with four or five benches" [for four or five men] on West 46th Street and started to manufacture. He hired two or three men for the polishing. When he knew that Shmiel and Aron were coming in 1941, he found an office on East 45th Street with twenty-five benches. Moishe described how the federal government at that time subsidized the fledgling U.S. diamond industry by allowing firm owners to pay five dollars a week per apprentice in order to teach them the trade while the government underwrote the rest of their salaries. The family firm participated for about a year and then dropped the program. Because many apprentices left to start their own businesses as soon as they knew enough of the trade to be independent, the family did not consider the enterprise worth the trouble.

In 1948, Aron and his three sons set up a separate firm nearby. Bonpapa was only minimally involved by then and he died in 1952. Shmiel and Moishe moved their office in 1952 to West 47th Street, where they remained. After Aron died in 1957, his three sons continued in their firm.

Aron had pioneered a method of polishing diamonds that differed from others. He believed that the stone had a uniquely optimum way for it to be processed. Each stone was to be understood on its own terms and was unlike any other. This had practical implications. Rather than allow the factory foreman to make decisions about how to cut the stone, he made the decisions. Both firms incorporated his philosophy in their shops.

Shmiel and Moishe are the two uncles I knew the best growing up. During my childhood they lived two blocks from each other in Queens. One hour away in suburban Connecticut, we went to New York often for family events related to rites of passage and Jewish holidays. Most of their children were older than my three younger siblings and I were, and I idolized them and loved being with them. The two uncles were self-educated and cosmopolitan. They attended university courses regularly, traveled often, and went to concerts, lectures, and museums. Their homes were filled with art, music, and books; conversation was erudite, passionate, and inflamed, usually about Jewish subjects that encompassed religion, ritual, Israel, history, ideas, philosophy, and politics.

We children were astonished to learn years later that my grandparents had tried to arrange our mother's marriage to someone in Antwerp when she was a teenager. My mother had fervently rejected this match, and my grandparents, independent in their Orthodoxy, were tolerant. My mother then fell in love with my father a few years after coming to the States. My American-born father, a physician and psychoanalyst, was a genuine outsider but inherently compatible in shared Eastern European/Russian roots and basic Jewish values. The family immediately accepted my father and his entire Boston-via-Russia family.

It fit that we did not live in New York with all of them. My mother had been a rebel in school in Antwerp as a child. She had been thrown out of four schools, often played hooky, and was always looking for adventure, everyone confirmed. It's not that school was so terrible or restrictive, she told us coyly; she just felt a pressing need to spice things up a bit. That was her explanation for having painted a pretty flower on the teacher's chair so that when she sat on the chair with her beautiful white dress, it would be just that more embellished and interesting to behold. This incident resulted in her father—the matter was too serious to simply call her mother—being summoned for a talk with Mme. La Directrice.

She rebelled against her older parents' Orthodox ways, their observance of *kashrut* [the kosher dietary laws of observant Jews] and their fussy protectiveness of her. She became an ardent Zionist and used to sneak out of the home for socialist meetings, she told us kids. Though she recalled anxious talk in her home every month about how to meet the mortgage payments, she despised that her parents were bourgeois capitalists, and she urged them to give all their money away.

Once in America she was a fundamental enigma to her older, more staid businessmen brothers and their wives, each of whom maintained European habits, continued to speak Yiddish along with English, and stayed fairly observant. Their English was painstakingly learned and forever carried the inflections of Europe, while her English was practically accent-free. She was a teenager, ruing the Europe she'd been torn from while eager to become American. Part of a different generation from her brothers', she came to the United States at a perfect time to assimilate American ways. Like much of the postwar world, she was entranced with America, yet she felt increasingly European the older she became.

The differences between her and her brothers were profound. Much younger than they, she was devastated about leaving her friends in Antwerp at a time in her life when friendships are the most intense and potential. Her brothers' adult concerns were the establishment of their business and their new families. My New York family's ways were in stark contrast to our Connecticut lives. Our mother was deeply ambivalent about her commitment to official Judaism, though her Jewish identity was not in doubt. A writer of short stories and children's books, a joint founder of one of the first suburban art galleries in 1960, she embraced the freer life available in the suburbs. Their intellectual and lively friends were artists, writers, psychoanalysts, and journalists.

Incredibly, many in the family had survived the war, and they would all come together for Passover Seders and other events over the years. We heard whispered accounts of survivor stories. We pointed furtively at the blue tattooed numbers on one of our relative's wrists and shivered with knowledge. Children ran around during these family times while the parents discussed and debated. There were a lot of children, and there was always ardent debate.

◆

Moishe and Shmiel always characterized themselves as the pessimist and the optimist. Though they agreed on the basics, Moishe was always concerned that dire consequences might occur, whereas Shmiel had a quiet confidence that a good outcome would prevail. In subjects concerning business, politics, or the fate of the Jews, they were similar in their outlooks, but their personal temperaments maintained this fundamental difference.

Shmiel seemed to me as a child a benign, quiet, tall, and protective presence. He had an understated and wry sense of humor. He used to make up stories for the children and tell them one-on-one in a gentle, conversational way. The stories captivated me with their exploration of tiny domains. I loved the wayward fly who defied his mother to explore forbidden territories, had frightening and lonely adventures, and finally found his tired way back to his mother again. The little penny found himself in dark pockets and rough hands and was slammed on counters in strange places but managed to survive somehow. Shmiel spoke in a quiet, low tone, his eyes enjoying the effect on us of his slowly unwinding story that he was making up as he went.

He was entertaining us, but he was just as much discovering us as he told his tale. He would gauge our reaction, calm our fears, reassure us that there was more, we should be patient, we should be trusting. He seemed to know me. My grandparents gave me a watch when I was 6, and I was proudly preoccupied with the competence and responsibility that owning this watch meant to me. Soon after I received it, Shmiel wrote me a letter asking me to tell him what time it was. I happily sat down to answer his question when I suddenly realized that I didn't know if he meant what time it was when he wrote the letter, what time it would be when he received my return letter, what time it was when I started to write, or what time it was when I finished the letter. These tormenting questions prevented me from answering. Many years later, it dawned on me that he was teasing me about my obsessive streak and introducing me to philosophy.

In his work, as I learned from observing him, he was watching and waiting, probing, listening, and creating. He was a discoverer, continuing the deliberate method his brother Aron had pioneered of exploring the individual stone's unique properties. He would search a rough stone's mysteries. He said that was what fascinated him about the work. The rough stone, looking like a dirty, uneven piece of unappealing quartz, had numerous puzzles and surprises hidden inside. His job was to strategize about the stone in hand. He was the primary one in the office who would place his discerning eye on the stone, identify its surface flaws, intuit its potential, predict the problems that lay within, and map out how to proceed. With deliberate patience he would sit, peering through the loupe at the gem-to-be, marking with thin black lines of India ink where the initial cuts should be so that further visual access could follow. A few years after his death, a diamond trader reminisced about Shmiel and said:

Shmiel was brilliant. He could tell what was of value in a pile of rough. He knew if you could make a profit or not. He could find the best way to make the stone. Everyone asked him for advice and he was so beautiful and kind. No one else in the office ever learned to do it like him, and when he died, that was the end of it. He was one of the most brilliant ever.

When the others were out of the office for hours at a time, he would scrutinize his babies in this manner, getting up to take a marked stone to the foreman of the factory, and to confer with him about a stone that was being worked on. My notes early in the research reflect what he told me:

About making the diamond, he's like a therapist, he tells me. He tries to bring the problems to the surface. While he's telling me this, he is gently fingering a rubber band, smoothing out its surfaces and flipping it over and over again. His big, delicate fingers extending long from pale hands large and blue-veined look like they're used to precision and taking great care. He looks for the imperfections. He marks where the stone should be cut. In that way, he has a "window" to see into the stone better. When he brings the imperfections to the surface, the patient feels better. He likes this part the best. When the stone is polished, he loses interest in it. He enjoys the challenge when the stone has problems.

With adults and with his children, I later learned, he was a domineering and fiercely opinionated man. He mined his conversation with argumentative traps, and he would pounce on what he considered a contradiction or flaw in the discussion. He had pet peeves regarding Judaism and many other subjects, usually because they would impinge on something Jewish. He was fiercely against any current intellectual fad. He challenged my father unceasingly about psychiatry and psychoanalysis even as he had deep respect for my father's professional and personal integrity. He challenged me, the anthropology major, about cultural relativism, which he thought defied all sense. He argued for standards, for an excellence that was intuitive and transcendent. Yes, one could try to understand other people and other ways, but one must never dilute or lower standards. He was against affirmative action. He was deeply grateful to the United States for having provided a safe new home for him and his family. He was against apologist arguments that would explain away any immigrant or native group from achievement. He rejected the idea that black Americans had any greater hardship than Jews in pursuing excellence and prosperity. He believed that all human beings had the existential and fundamental challenge of trying to justify their lives, make a living, and make a contribution. He believed that in this respect all human beings were equal and suffered the shared burden and privilege of making their lives worthwhile. Our arguments were hot and lively.

He seemed bemused when I pitched my "diamonds project" to him in 1986. Skeptical and a little curious about what I had in mind, he put me off and then seemed to encourage me to pursue it, ultimately agreeing that it was a worthy subject. He seemed personally gratified that he and I would be getting to know each other better.

Shmiel was always the "inside man" of the family firm. More shy and retiring than his younger brother, Shmiel held himself somewhat apart, observing and assessing from a slight distance. He spent many hours at the desk in the office. The first hour or so he would converse with his brother in Yiddish. As the firm expanded with one son each joining them, these men were included in the discussions that alternated between Yiddish and English. They would go over yesterday's business and today's plans, comparing notes. At 9:30 or 10 A.M., one or more of the men would go to the club or to offices where they had made appointments or promised to show or to pick up goods. Shmiel would stay behind, working on the rough stones, conferring with the men in the shop, and taking phone calls. Later in the morning he would go downstairs, cross the street, and go to the club for lively conversation and deal possibilities, joining the others. Standing up to answer a phone, he would grab someone who might have a stone they were interested in, or look for someone who should give a stone back so it could be shown to someone else. Sitting separately or with others, he was busy talking, watching, showing.

At lunchtime the brothers and their sons would usually head back to the office and have their sandwiches and fruit, answer phone messages, place calls, check with one another about what was happening. Shmiel would check on progress in the shop. He'd look over the stones, confer with Moishe, son, and nephew if there was a question, and give a stone back for more polishing on the wheel. After lunch Shmiel would often stay in the office to continue his work, though sometimes he would go back to the club. In the initial phase of my research, he was turning 80 and his normal workday was approximately 9 A.M. to 3 or 3:30 P.M., plus the half-hour commute on the subway and the half-hour walk from subway to home. He was able to continue work until shortly before he died at age 85.

Moishe, the youngest brother, died in February 2000 at age 88. He worked until the day before he had surgery, the complications of which resulted in his death two months later. Diamonds represent excellence, he told me early in the research, and excellence is rare. The pursuit of excellence for both brothers was intense and constant. Judgment could be severe. Like his brother Shmiel, Moishe strictly maintained standards for himself and for others. Never suffering fools, he had withering disdain for mediocrity and inferiority and could be very stern. An arbitrator for years, he was respected and sought out for his fairness and honesty. He enjoyed the puzzles of arbitrating disputes much as Shmiel liked the puzzles tenaciously deep in the diamond. Scorning politics, he never held an office in the DDC. His blunt honesty shortcut tact, but the light that sometimes flickered in his eyes lightened the sharpness. Once, when someone was pleading with him to consider selling a stone for his price, he faced the man, took both his cheeks between thumb and forefinger, gave an affectionate squeeze, and, pressing nose to nose, said, "No."

Another incident reflected this same flavor:

We are about to go in the club with Mike, a friend of Moishe's, when Mike is confronted by the guard brusquely and told he can't go in. The guard repeats that the man can't go in several times and bars his way as well. Mike protests

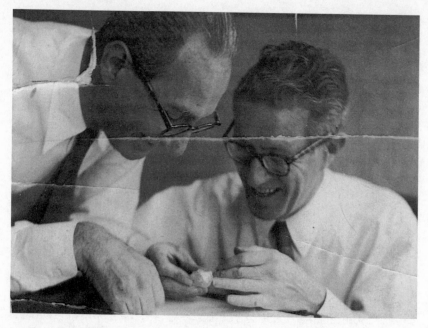

Moishe, left, and Shmiel are pleased with the high quality of the stone they are about to cut, circa 1960. (Author's collection)

that he heard him and all right, he's not going in. We enter and Moishe asks one of the club officials why Mike was barred. He hadn't paid his dues. He was sent two notices. Moishe persuades the official to let him in this once; he'll vouch for him; he'll pay. Mike is let in. He is fuming, irate, completely ranting. He's been a member for forty years! He can't get his mail! He was in Europe, that's why he didn't get the notices and couldn't pay his dues! He had to take care of his mother! Did we see that, how he was treated? They wouldn't let him in! Moishe elbows me away from him, says I shouldn't get a bad impression from him. Mike keeps up his tirade. He says to Moishe: What would *you* do? What would you *do!* Moishe slowly turns to Mike, serenely places his hands on Mike's shoulders, and says barely above a whisper, "I would shut up."

Tall and erect, Moishe had obvious appeal to young and old, men and women. Dressing meticulously, wearing subtle colors and taking care to match tie, shirt, and jacket, he had replaced formal, uncomfortable shoes with black running shoes in recent years so that he could be on his feet. Moishe was always adventurous and lively. Enjoying people, he relished good jokes and savored gossip and personal stories. This trait made him the natural "outside man" for the business from the start, when the family was still in Antwerp. Bold in a respectful but interested way, he always found it easy to approach people and to converse with them. Though meditative like

Shmiel, he always thrived on the company of others with whom to talk, joke, argue, and compare stories. The most public of the brothers, Moishe was the most visible.

These two uncles and their firm were respected for their toughness, honor, honesty, and integrity. They let me explore this rich world.

[4]

New York, Diamond Center

"The amount of merchandise on 47th Street is mind boggling. There are more jewelry resources, and buyers and sellers on that one block than anywhere else in the United States. When it comes to jewelry, if it exists you can find it on 47th Street, or someone will know where to get it, no matter how obscure" (Michelle 1997).

The United States is the largest diamond market in the world, and New York is the place through which most of this huge supply travels. New York is also considered the most skilled cutting center. Though India and Israel dominate the manufacture of the smallest diamonds, and Antwerp is the major distribution center, New York is known for its manufacture of the finest and largest stones, of 2 or more carats. New York has some of the best cutters in the world. "It is uncommon, but not unheard of" says one trade writer, "for these cutters to produce finished stones of more than one hundred carats . . . the 'crème de la crème'" (Blauer 2001:40).

In fact, the Diamond and Jewelry Industries Study conducted by the Office of the Manhattan Borough President reported that 95 percent of diamonds that are imported into the United States pass through New York. The industry employed more than 23,000 people in New York in 1991, which constituted approximately 25 percent of the national employment in the jewelry and diamond industries. In 1998, the U.S. market had $7.9 billion worth of polished imports (Rapaport 1999). The New York diamond district includes about 1,800 licensed dealers. The strength of the city's diamond industry has always been in its high quality, its yield, and its ability to specialize (Shor 1997a:64). New York has about 100 mostly independent

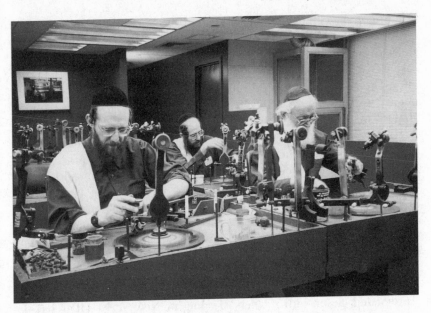

Cutters at work at the wheel. (Photo by author)

manufacturing companies with a total of about 600 cutters (Global Witness 2000). Though Japan grew exponentially in the postwar years in its consumption of diamonds to become the second largest market in the world, it was still surpassed by the United States. In fact, U.S. retail sales grew 2¾ times from 1982 to 1997 (ADIA, spring 1997:1), and 1998 sales totaled $22 billion (Global Witness 2000).

Among those in the New York industry, a greater concern than competition from Japan has been the effect of commoditization, the global economy, the Asian financial crisis of the mid-'90s, the economic slowdown, and ongoing fears of terrorism in 2001. As De Beers announced its monumental decision to privatize and to market its jewelry in its own stores in 2001, the belief among New York dealers and brokers that De Beers does not support the New York trade intensified. In addition, manufacturers worked to create new patented cuts as a way to beef up marketing and sales (Weber 2001).

New York diamond dealers and brokers have felt the stiffer competition of the global economy. While the CSO has stressed the strength of the New York district (Shor 1997a:64), a chronic shortage of rough goods plagues the New York industry. Though the number of New York manufacturers has decreased significantly over the last few decades because of lower wages in third world countries (Myerson 1992), New York remains the manufacturing center for larger and finer stones. The strong reputation of the "New York make" has been preserved as cutters have tried to achieve the most beautiful stone with the least sacrifice to the weight of the stone. New York guards its territory as best it can as the world proliferates with diamonds and with cutting and marketing centers (Shor 1997b,c; Parker 1999). Political jockeying and lobbying are essential in the trade. For example, lobbying efforts by New York diamond officials persuaded the GIA not to open a branch in Antwerp, to maintain New York's importance in that sphere ("DDC President Advises Members to Specialize," 1995).

The high standard of manufacturing for which New York is known seems to be in increasing demand worldwide, especially in East Asia. As consumers become more educated about diamonds, they demand a higher standard, and this trend works in New York's favor. Because general profitability has decreased with the growing commoditization of diamonds, however, fewer foreign companies have seemed willing to undertake the trouble and expense of setting up business in New York. Already established New York businesses are reassured by this.

With the number of manufacturers down, brokers and other middlemen face an increasingly uncertain fate. Meanwhile, manufacturers and dealers are working closer with retailers. Trading in New York's DDC has decreased markedly, but trading is down in bourses worldwide as more people trade in individual offices. New York trade leaders instituted "Market Weeks" at the bourse in the late 1990s to drum up business, and other efforts have sought to increase the visibility of the New York trade and to highlight New York's preeminence. Enterprising individuals have scouted specialty niches, and some manufacturers have expanded into smaller goods because their experience, fast service, and specialized knowledge of the U.S. market could offset the higher wage levels. Top diamond companies, such as M. Fabrikant and Lazare Kaplan, and smaller companies have developed finished jewelry lines for their loose rough to capitalize on the higher markups possible in these goods. Others have gone into design. Some companies have considered using automation and computer technology to manufacture smaller goods, to be more competitive. By 2001 some large manufacturers and sight-holders had acquired cutting and distribution companies instead of contracting with them (Weber 2001).

In all, New York seemed to be maintaining its overall symbolic and economic preeminence in the world diamond arena at the beginning of the new millennium, and personal philosophies seemed to rule the perspectives of individual traders. Perennial pessimists said that the business had never been more miserable and that their children should stay far away from it. Other, more contemplative, souls were perhaps better cushioned financially or temperamentally. More likely to shrug and counsel that the diamond business has always been hard, they added that people either find a way through the difficulties, or they don't make it.

The foreman of the shop among his tools. (Photo by author)

◆

In America immigrants' assimilation to mainstream life is not a straight upward progression but is patchwork and zigzag. Family members are vital and trustworthy employees in that process, willing to work hard. As Zenner writes, "The family is the main welfare and employment agency for individuals who are not supported by the state" (1991:67). Mitchell (1978) describes how Jewish family clubs in New York helped support immigrant members financially and socially and overlapped with joint businesses. Conflict as well as solidarity resulted from this conjoined role (cf. Leichter and Mitchell 1978). "If a family member has 'the credential,' the network can assist him or her in achieving both individual and family goals," wrote Silverman (1988:165) about a multigenerational Jewish family in Pittsburgh.

Jews in America found niches in the ready-to-wear garment trade, mass communications, and department stores; they went into businesses serving the poor where there was a demand (such as pawnbroking); and they persisted in previous trades, such as innkeeping or tailoring. Employment of Jewish immigrants by Jewish employers entailed the trio of low wages, low search costs, and low entrance barriers (Kahan 1986:105). Morawska (1996) reported that when Jews immigrated to smaller cities in the United States, such as Johnstown, Pennsylvania, they often stayed in the same kinds of small-scale business enterprises in which they had worked previously. The Jewish immigrant brought commercial skills and international connections that helped some businesses, such as the diamond trade. While most Jews departed the middleman niches in favor of professions and corporate business in the latter half of the twentieth century, some Jews have maintained the viability of Jewish-concentrated businesses such as the diamond trade, where large numbers of separatist Hasidic Jews have found refuge (see Zenner 1988). 47th Street is one such place.

For many years an African-American security guard stood at the entrance of the Diamond Dealers Club at the corner of 5th Avenue, where he would greet the diamond traders by name. On Fridays as they were leaving, he would wish a good *Shabbes* to them. He knew which of the Hasidic Jews were of which sect, and he would tell them the exact time that the Sabbath candles were supposed to be lit. As these long-bearded men strode out of the club in their street-sweeping black caftans and black hats, I wrote in my notes what the guard told me:

The Lubavitchers and the Satmars don't get along, you see. This week one group lights the Sabbath candles at 4:36, the other at 4:39.

As people left the club, he also reminded them of the name of the weekly Sabbath portion that would be recited that week. They thanked him for the perhaps unnecessary reminder.

The New York diamond trade is a great ethnic mix within the cosmopolitan smorgasbord that defines the city. As Kirshenblatt-Gimblett has written, "Group boundaries are not 'given.' Rather, they are socially constructed and situated, constantly negotiated; they are multiple and complex" (1987:89). Like other groups in the United States who do not assimilate into the dominant culture, Jews have not necessarily been open to communities in which they reside. Though the 47th Street diamond district today is more than 95 percent Jewish and seems homogeneous to outsiders, it encompasses ethnic diversity, degrees of religiosity, as well as varied individual and group histories. It enmeshes itself in the city and yet remains separate. Most of the older Jews who work in the district originally came from Antwerp, Amsterdam, London, and other Western European towns and cities, and many of them had recent ancestry from Central and Eastern Europe. Today many Jews in the business are the American-born sons and daughters, nephews, nieces, and grandchildren of those original European Jews. As do American Jews in general, they occupy a gamut of religious observance and affiliation (see also Freedman 2000, Heilman 1982, and Heilman and Cohen 1989).

Other groups of Jews come primarily from Iran and Israel, with fewer deriving from India. Some of them immigrated to the United States recently, whereas some immigrated one or more generations ago. These Jews also vary in their level of Jewish observance, though the Israelis are by and large the most secular and least observant among them.

The Hasidim are a relatively new factor in the diamond business in New York, but because they are so visible and their numbers have grown so quickly, they are almost synonymous with the business to the outside world. The American-born Jews in the diamond business are increasingly Hasidic.

Hasidism began in Eastern and Central Europe two centuries ago to revitalize Orthodox Jewish belief. It was a religious response to modernism and secularism. It challenged traditional Orthodoxy. Hasidism had a resurgence in this country around World War II and is a strong force in Israeli and American Jewish society today. Joselit has written about its impact among New York's Jews: "Challenging the indigenous, happily modernized Orthodox community at its very core, the newer element disdained the education of the New York-trained rabbi, mocked the community's preoccupation with manners as *narishkeyt* [silliness], and spurned its religious culture as ill-conceived, inconsistent, and most damning of all, inauthentic. Virtually everything championed by the interwar Orthodox, from English-speaking rabbis to 'modern yeshivas,' was anathema to the emerging right wing. In due time, not surprisingly, a social chasm developed between the two groups by mutual consent" (1990:147).

Mintz estimated that Orthodox Jews represent about 8–10 percent of American Jews. They consist of the strictly Orthodox (about 5 percent) and the modern Orthodox, many of whom do not wear *yarmelkes* or distinctive clothing (1992:6). Because the Hasidim wear clothing that sets them apart from mainstream North America, outsiders often lump together the individual sects (or courts) though they encompass different groups with varied dress, prayer observance, and rituals, including calculating different times to light the *Shabbes* candles. These differences are "expressions of a cosmology and theology that go far beyond dances, songs, and stories" (El-Or 1994:12). Though sharing a common history, they fervently adhere to different interpretations of the law and revere separate spiritual leaders, *rebbes*, who they believe can heal and bless.

Like fundamentalist groups such as the Amish and others, the Hasidim deliberately set themselves apart from the secular community, which they consider polluting and harmful to the fulfillment of their spiritual quest as Jews. Their daily lives are governed by the 613 Jewish commandments (*mitzvot*), which determine their lives of piety and devotion to their *rebbe* and God and affect how and when they pray, what they wear, and what they eat.

Hasidim often live and work together in environments that respect their religious devotion and customs; this helps explain the attraction many Hasidim have to the diamond business and why the numbers of Hasidim in the business has burgeoned in the last few decades. The industry allows individual variation in how its members practice the trade, the environment is

Jewish, the trading club provides a *shul* and kosher eating premises, and most business dealings are with other Jews. Offices and the club are closed on Jewish holidays and the Sabbath. Since most traders are male, Hasidim are spared doing face-to-face business with women, with whom they are prohibited from contact (other than their wives); however, some choose to conduct business with female traders. Most Hasidim speak Yiddish as their first language, and Yiddish is a standard language of the trade.

Kranzler estimated that approximately 15 percent of the younger and middle-aged Hasidic workers in the Williamsburg section of Brooklyn were employed in some aspect of the diamond, jewelry, gold, and silverware business (1995:193). Families there had an average of seven to nine children. Many of the male children became cutters, brokers, and dealers, while many of the women worked in secretarial positions in diamond offices. A trader I met had ten children, the oldest of whom announced his engagement soon after his ninth sibling was born. Many Hasidim could barely support their large families and were not always able to survive in the business.

Hasidim and non-Hasidim within the business did not always get along. The beliefs about the Hasidim by the non-Hasidim revealed some of the strains between them. Some non-Hasidim said they considered Hasidim uneducated, even illiterate. One said they were barely conversant in English because they've grown up with Yiddish. He criticized the narrowness and provinciality of their world. A dealer told me that the Hasidim expect the work to be easy. He said that Hasidic children are mediocre, don't work hard, and are "dumped on the business and cannot make it" elsewhere. They were sometimes disparaged for their large families. Some non-Hasidim linked the prolific Hasidic birth rate to their "taking over" the diamond industry. Some non-Hasidim said the Hasidim had a "mob mentality," gave the business a "bad name," and kept it from advancing into the "modern" world. Hasidim, for their part, were aware of and resented these disparaging attitudes. They rejected the less observant or nonobservant behaviors of other Jews. Many divisions obviously separate groups of Jews.

In addition to the predominance of Jews in the diamond business, the last few decades have seen the introduction into the industry of traders from East Asia. With the significant increase in diamond purchases by Asian consumers, larger numbers of brokers, buyers, manufacturers, and other traders are from China, Malaysia, Korea, Japan, Thailand, and elsewhere in the East.

As these groups from the East and the West deal with one another, create stereotypes about one another, and make various concessions to trading with one another, they accommodate one another. The most striking symbol of their accommodation is that no matter the ethnicity or nationality, the words that close a business transaction are always the Yiddish *mazal und*

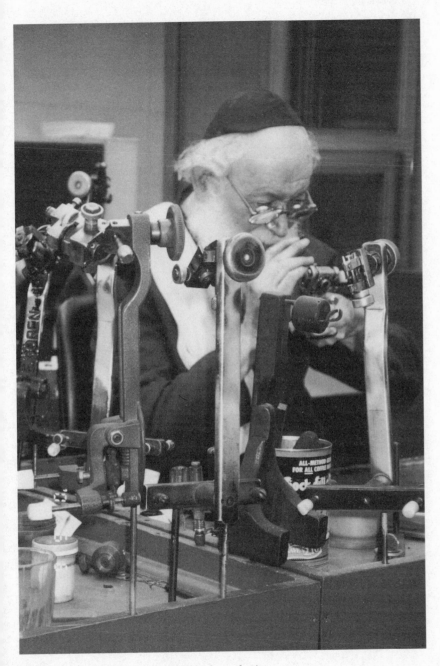

Hasidic cutter examines a stone. (Photo by author)

brucha or simply, *mazal* [luck and blessing, or luck]. The differences exist beneath the flag of the Yiddish saying to which all adhere, and the constant ethnic changes of the district illustrate both continuity and flux.

◆ ◆ ◆

[5]

Million Dollar Handshakes

"Well, I don't know if I told you the story of the young man who came from Poland," the manufacturer began. "He had an uncle in Antwerp who was a diamond dealer. It's a wonderful story. Let me see how it goes." The manufacturer settled more comfortably in his chair, readying himself for the delicious tale he was about to relate. "He came to his uncle that he would like to start in the diamond business. But he needed a 'lot,' a parcel paper of small diamonds. That's called a business. Literally in Yiddish, it's *geshceft*. So the uncle said, 'Why not?' And the assumption was that it would be on credit because he didn't have any money. So he bought this business of 10 carats and let's say he has 90 days to pay for it. So he takes the goods and he looks at them, he studies them and he has ideas how to sort the goods—he tries various things." Here he used his hands to show how the hero of his story attempted—with little success—to group the stones this way and then that way. "He tries an assortment and puts them on the market, but no matter what he does, he sees he paid too much for the goods. It's a sheet that when you cover the face, the toes show, you know. So he's not very happy. He's got to return the goods, but how does he do it? He's got a cousin who's a nephew of the same uncle, and he tells him he can't make out in Antwerp, it's too difficult a life, too difficult to do business, and he's going to go to New York."

The manufacturer paused, looked at me: Was I still paying attention? His look signaled me that the story was getting good now. "So the cousin runs right away to the uncle, and he says, 'Listen, Uncle, did you hear this nephew of yours is going to New York? This is what I just heard.' Says, 'Uh-oh.' [Maybe he was going to abscond with the goods and not pay him?] 'Tell your cousin that I want to see him.' So the nephew goes over to the one from Poland and tells him that the uncle wants to see him. So OK, he goes to the uncle, says, 'What is it? Why do you want to see me?' Says, 'Yes, I was just curious. How did you make out with those goods I sold you?' 'Oh, I did very well, very well.' Says, 'I sorted them and they look very good to me.' The uncle says, 'Maybe you would like to sell them. You still have them?' Says, 'Yeah, I still have them.' Says, 'I give you this and this much profit for it.' So the guy says, 'I can't sell it for that price.' 'OK, I'll give you another guilder, make it a 2 percent profit.' Says, 'You've got a deal. *Mazal u' brucha.*' Returns the goods and the guy gives him the difference. Then the guy says goodbye to him. The uncle says, 'What's this I hear, you're going to New York?'

"'Go to New York? If I can make a profit here?'

"I think it's maybe a true story. It's probably true. A good place for stories, Antwerp."

◆

The manufacturer who told me this story wanted to illustrate the competitive flavor of the diamond business. This competitive spirit exists even— or maybe more so—within families. Though this story is old, the business is, if anything, more difficult today. This chapter takes us into the New York marketplace of the diamond business for a look at how deals are made.

"The very nature of the trade encloses the world of diamonds in an aura of secrecy," wrote Legrand. "No one outside the profession is admitted to these halls, and it is not easy to become a member of the exchange. Each is a closed corporate society governed by its own laws. Members have a deep sense of allegiance to the society, and customs and traditions are strictly observed. Prospective members must provide solid proof of many years of successful and irreproachable business activity, and each candidate must be proposed by several members of the corporation. Once admitted, members answer only to their peers. Disputes are never taken to the civil courts, for each exchange has its own tribunal whose decisions are final. The penalty for refusing to abide by these decisions is permanent exclusion from the profession throughout the world. Total integrity of conduct is assumed" (1980:234).

Thus does the author introduce us to the world of the diamond bourse. There are twenty of these in the world, including four in Antwerp, one in New York (there used to be two), and others in Tokyo, Tel Aviv, Mumbai (India), Hong Kong, Singapore, and elsewhere. New bourses in Thailand, Japan, and Florida have recently applied for membership to the WFDB, reflecting their major activity in the diamond industry. Each WFDB member

Diamond lamppost at entrance to 47th Street, "Jewelry Way." (Photo by author)

is structured similarly. Each acts independently at the same time that it co-ordinates with the world umbrella body. The WFDB structure ensures enforcement of uniform trading and ethics rules so that if a trader is expelled from one club because of unethical business practices, he or she is denied access to the entire network. The excommunication has profound practical as well as symbolic implications.

The WFDB was founded in 1947 to facilitate such uniformity among the worldwide bourses. Distressed at the small amounts of rough that De Beers was supplying to New York dealers, the president of the New York DDC met with the heads of several European bourses to present a united front to De Beers (Lubin 1982:47).

The bourses are the symbolic centers of the diamond business, the emblems of the diamond marketplace. Street activity, the business conducted in the booths in the exchanges that line 47th Street, contrasts with the business upstairs. One dealer referred to the street trade: "It's another world down there; they are not us." Another summed up: "They're beneath us." The club, ten floors up, is peopled by the smaller-business men and the foreign buyers sponsored by local members.

The club contains people who are well educated and those who are not. The many nationalities create a buzz of languages lively with difference. As I was finding my way in the club, I noticed the *shul* and the sixty-six volumes of the Talmud lined up on shelves, splayed open on tables. Some people in the DDC know all of the Talmud by heart, a trader commented—though they might know little else, he added slyly. This trader knew a member who had graduated from the Sorbonne in physics, and someone else who had probably never been north of 57th from his daily Brooklyn to 47th Street commute. The diamond business can be an entirely self-enclosed world or a living that allows varied outside interests. A few current club officials hold Ph.D.s.

The transmittal of information in the club helps make up the cacophony. One gets a feel of the marketplace. A person is seen, mentally noted, remembered. Simone Lipschitz's fond recollections of childhood visits to the bourse in Amsterdam are easy to picture in the modern-day New York DDC: "There was a cosy atmosphere at the Exchange: business was not separate from everyday life. A sale, which might have involved lots of bargaining, was invariably followed by a homely 'mazzel en broches.' . . . Some members sat calmly at tables while others used to rush around trying to make an elusive and precarious living. There was a continuous hum of voices in the hall, and every now and then a raised voice would break the surface. . . . I felt at home. . . . The activity on the Exchange floor reminded me of the movement of a lake or of the sea: swirling currents, ripples and waves. Suddenly part of the floor might stand bare and empty, and further along there would be throng of people jostling. An argument? News affecting the diamond trade or politics?" (1990:15).

The DDC in New York is a nonprofit organization that includes a taxable corporation for the trading portion of the club. The club operates on a cost-plus basis, with members pooling their dues to maintain the expenses and services of the building, the telephone and security systems, and the like (Lubin 1982:36). Until recently the DDC was a hive of fervent activity, crammed full of men and a few women at the tables and in the aisles, examining the stones and making trades. This activity equaled if not exceeded the business conducted in the private offices, which has increased in recent years, dwarfing the club in comparison.

One observer wrote in 1981: "The ceaseless round of activity [in the DDC] suggests that all diamond trading takes place there. In reality, more transactions are going forward in the privacy of offices up and down the street. 'Important buyers go from office to office, not to the Club,' explained [past] President William Goldberg. 'In the main, the Club is for the small man, who will sit there and look for business. But it is more useful for the foreign buyer who doesn't have an office of his own. It's the check-in point for him, the place to be seen and to make contact. And the word goes out, 'So-and-So is in town; call him at the Club'" (Green 1981:201).

DDC rules stipulate that a prospective member must have been working in the diamond and/or precious stones trade for at least two years and must write an application endorsed by four current members (with exceptions for children and spouses of deceased members). An admissions committee evaluates the prospective member, and the DDC board votes on the recommendation. Before admission is granted, the name and photo of the applicant is posted in clubrooms around the world, allowing members to voice objections. New members are on probation for two years. Members for twenty years or more may receive a reduction or waiver of dues. Special provisions in the DDC bylaws determine the process when members fall into debt or declare bankruptcy. After a suspension, a committee meets to devise a method for the debtor to pay his creditors that, if properly fulfilled, can allow the debtor to remain in business.

Lubin relates the story of an extraordinary settlement after a member was arrested and threatened with jail when he lost his money in the stock market and reneged on his debts: "[We] had received word from Antwerp that the debtor's family was willing to put up $200,000 in rough diamonds as payment to the creditors, if the debtor was kept out of jail. We went to court together, and met with the judge in his chambers to explain the situation. Although our creditors wanted the defendant punished, we told him, most of them were small-business men, and would be better served by receiving at least some percentage of the money owed, so that they could continue in business. The district attorney . . . vehemently object[ed] to any light sentence. . . . The judge decided to sentence the man to 2–5 years in Sing Sing, suspended the operation of the sentence, placed the debtor on probation, and ordered him to pay the $200,000 to his creditors within sixty days or the sentence would be carried out. The $200,000 came from Antwerp and was distributed to the creditors" (1982:34). When I showed this quote to a dealer during field research, he exclaimed, "Way in the past!"

Because every member agrees to abide by the trading rules of the club, the international rules help diamond traders conduct business and resolve internal disputes rather than appeal to outside jurisdiction. Reasons for suspension may include non-payment of dues, making untrue claims, filing for bankruptcy, not meeting business obligations, not adhering to an arbitration decision, or avoiding arbitration by going to court against another member.

The DDC provides a protected, common place for trading. Social and health services for needy members are provided. A trade magazine called *New York Diamonds* is distributed to all members. The club management meets with local and international officials to support the interests of the industry.

The New York DDC was founded in 1931 in the midst of the Depression by twelve men who decided it was to their advantage to find a safe and central trading place. Apparently, a dealer had dropped and lost a diamond on the street while examining it, a common practice at the time (he found it later that night in the cuff of his pants), dramatizing the need for a safe, indoor trading place (Lubin 1982). Clubs in Europe provided the model for New York. The first New York club, of fifty members, was located in the Bowery.

A jeweler in his 80s related why the club was first located there. At 13, his job was to buy cameos for the diamonds in them for his father's manufacturing business. "It still hurts me to think of it," he told me. "I was breaking up the cameos and popping out the diamonds, taking the diamonds and shipping them to my uncles in Holland. They were small—we never cut small diamonds in this country—so they were cutting them in Holland. This way my father's brothers had some work, because they were completely destitute in Holland in those days because of the Depression. I was sending them the diamonds, they were cutting it, sending it back to me and getting whatever the labor was, and whatever profit I had, my father let me keep. . . . People realized that you could get secondhand jewelry down there [in the Bowery], and the exchanges started around 1930, give or take, in the Depression, and there were some people down there who did very well."

During World War II, the influx of immigrant diamond traders, especially

Hasidim discussing business on 47th Street. (Photo by author)

from Antwerp, outgrew the club space and prompted a move to West 47th Street in 1941. The club on 47th Street established a formal jewelry presence, reflecting the jewelry firms that had begun to locate there. The jeweler who had removed diamonds from cameos moved uptown when others did. "Actually," he continued, "48th Street was the main street. Rockefeller Center was being built and was just finished, and the building 608 Rockefeller Center was a beautiful building, and all the 'better' firms moved into that building. All the diamond dealers, because they gave tremendous deals: they would give a year free rent just to get the people in. They would buy up all the leases. There was plenty of money there from the Rockefellers, and the dealers took advantage of that, and that became a diamond building. So the surrounding area, 48th Street, became the factories. (No manufacturing occurred in Rockefeller Center.) And I asked my father, 'Dad, all our customers are now uptown.' And my father said, 'But all the working people are downtown.' You see, diamond cutting, you need people to scour the wheels. I said, 'They'll come uptown.' ... And my mother agreed with me. And we moved uptown, must have been '39 or '40."

Lubin described the new 1941 club like this: "Tremendous, bare windows and high ceilings exaggerate an unavoidable sense of bedlam. ... A good part of the bedlam comes from the paging system, which consists of 46 ceiling speakers that are in use most of the time. Despite the dull roar, each dealer seems to hear his or her own name when it is announced. ... Another part of the bedlam springs from merging of hundreds of voices in at least a half-dozen languages" (1982:18).

Over the next decades the club again outgrew its space. After rejecting a move to 43rd Street in the 1970s, in 1985 the DDC renovated a corner building on West 47th Street that fronted on 5th Avenue. In the next few years, the GIA, a jewelry security firm, a post office, and customs clearance center for jewelry imports moved into the premises. Approximately three hundred diamond-related tenants occupy the building. Though more services, offices, and the club helped to centralize the business in New York, great movement in and out of the building still occurs as traders travel to other offices, banks, and jewelry firms throughout the district.

◆

Trading in the diamond business follows rules that have been observed for generations throughout the world. According to the DDC bylaws (1980), "Any oral offer is binding among dealers, when agreement is expressed by the accepted words 'Mazel and Broche' or any other words expressing the words of accord." Young and old dealers alike observe the oral form of agreement and consider it trustworthy.

Buyers and sellers, often with the aid of a broker who receives a 2 percent commission on the sale, negotiate the selling price of a particular gemstone or group of stones (known as a lot). The diamond is contained in a small, elaborately folded "parcel" paper. Usually the trader carries numerous parcel papers in an expandable wallet, or less typically, a briefcase. Because the U.S. trade concentrates on individual stones for particular customers, a potential buyer often initiates the procedure by asking if the seller has a certain kind of stone. For example, he might ask, "Do you have a 4- or 5-carat, J-K, pear shape?" Alternatively, traders show one another their goods in the hope that something might be of interest. I sat at a table at the DDC and watched some of the goings-on: Two men sit across from each other. One tosses a parcel paper across the table to the other. The other flicks it open, first fold with thumb, next fold with thumb and forefinger, looks briefly, folds it instantly back up, and tosses it back. The parcel has an inner glassine paper, bluish in tint, in which the diamond is nestled. Another parcel paper sails in front of him with the same results. He looks up, asks for a particular size, clarity and cut, catches another parcel, looks quickly with loupe, says he doesn't like the cut; as he slides it back, another seamlessly lands in its place for perusal.

As the traders talk of price and terms, they refer to the Rapaport price sheet, a widely accepted listing of wholesale diamond prices for polished stones of different cuts against which standard discounts are applied. "I'll give you forty below" [the list price] is countered with an offer involving the length of time required to pay, such as, "thirty-eight below and sixty days." The delay in payment is an important form of credit. They go back and forth. Conversation is interjected. Another person comes over, and they start to *schmooze*. Health. Latest movies. Someone's bypass. Weight training. Then: "Oh, by the way, do you happen to still have that L 4-carater VVS1 that I worked before? I might have someone." The focus shifts slightly to the new person as the original trade is momentarily sidelined. "No, it's working with someone else, but I'll see what I can do. Maybe I'll get it back." Phone call to check, then back to the table. New people have joined. More parcels are passed. Someone recommends a Chinese restaurant. Another person says, "I was asked the other day if I had a good Chinese, and I thought he was referring to a new cut. He wanted a Chinese restaurant!" A few laugh. They carry on, continuing examining and talking. Several studious ones take turns peering intently through loupe and exchanging parcels with others, alternating conversation and bids.

If an offer is accepted, the parcel paper holding the gem is placed in a small manila envelope and sealed, usually with the buyer's name written across the seal. The terms of the offer are written on the envelope, and the offer is considered binding until 1 P.M. the next business day or another agreed-to time. This sealed offer is called the closed cachet, derived from the French word for hiding, a reference to how easily diamonds can be hidden.[1] The cachet is taken to the official weighing office at the DDC and weighed. If the seller agrees to the price then, the two traders say the traditional Hebrew words "*Mazal*" or "*Mazal und broche*," meaning luck and blessing. Usually a handshake accompanies the transaction. When deals are conducted in offices, a bill of sale is usually written. "Is it *mazal*? Do we have *mazal*? Yes? *Nu*? [Well?] *Mazal*!" Though the traditional deal is more commonly made face-to-face, I heard *mazal* said over the phone too. The word *mazal* seals the terms that have been agreed to orally, and what's written on the envelope confirms that.

"He couldn't just give me *mazal* immediately," one trader told me, "and finally, after he was agonizing over the decision after two to three days, he finally came to me. 'OK, *mazal*, you can have it for the price I offered.'"

Sometimes I saw deals consummated without the envelope-sealing ritual. This example illustrates that exception: Two old friends, who have traded together for more than forty years, are considering a stone they've discussed before. Ari is eager to sell the stone at 5 (hundred dollars per carat), but Zvi

---

[1] An informant was learning Japanese and used the Japanese word for "sealing" an item, hoping that would convey his intent to have the cachet done. The Japanese trader laughed at him and said in Japanese, "No, you 'cachet' it." Whenever they make a deal with this informant now, they tease him by saying, "Yeah, we'll 'close' it," using the Japanese word that is generally used for closing something like a window or door.

is rooted at 3. Zvi takes another brief look at the stone. He's walked away from this twice, but they both want a deal. They're impatient and want it finished. Ari says, all right, you can forget about the points (meaning the fraction of a carat that it's over in weight). In effect, he's coming down to $360 or $370. Zvi doesn't say anything, looks at him intently, then slowly offers his hand. Ari says *mazal*.

There are some unspoken rules of trade, as well. Breaking these rules alienates people. A dealer resented another trader's behavior. "He has been known to step on a few toes in the process," he told me. "When one is offering on another's stone, there is an unspoken etiquette. I'll give some examples. When there is an offer on a stone, aka the stone is in cachet, even if the stone is not technically sealed in an envelope, it should not be shown to anyone other than the buyer or maybe the owner. Now, I don't think he broke this rule. But another unspoken rule is you don't sell someone out for a minuscule amount. I've seen him do this a few times. Someone offers him on a big business [a number of stones] and he refuses the offer without giving a counteroffer. This is taken to mean he wants much more, maybe 5 percent more. Then he'll sell to someone else for half a percent more. This is how he burned a few people, including [us]."

Most of the time, of course, the stones that are shown do not result in a sale. Traders sit across from one another at the tables and show parcel after parcel, then jump up and go to another trader at another table and start the process again. They wander from table to table, restless, irritable, looking. They are interrupted by traders and phone calls inquiring about particular cuts and sizes.

Traders share information but they also guard their information. Guarding precious knowledge can give an individual an edge in business. When you know a buyer's preferences or what a person is likely to need, you have an advantage over people who don't have the information. Knowledge is power. Sharing it can diminish it. I was told an old story from Antwerp.

"OK, Rosenbojm, what was his first name—Yankel, that was his first name. He was an old dealer. He was a son-in-law of Finkelstein, who was a legend in the trade in the '30s and '40s. He was a joker. He used to say until the age of 50 a young man has to seek to make a living. After the age of 50 he has to see that the other one does not make a living."

Competition has its special sting and pleasure.

How traders look at stones is distinctive. The dealer's delicate, thin tweezers descend like insect antennae to separate small stones skittering on the parcel paper. He brings a tiny gem to his loupe to look intently at it. His body bent over the task looks as if he were scooping lo mein into his mouth with chopsticks. A man nimbly takes a 3-carat stone and places it tenderly and gingerly between his third and fourth fingers clamped

Traders in the Diamond Dealers Club on a typical day. The traders' faces have been blurred for security reasons. (Photo courtesy of the Diamond Dealers Club of New York)

firmly together palm-side down. Surrounded by black, wiry hairs jutting crazily from the man's hand, the naked diamond glitters as the hand is turned this way and that to pick up nuances of light and to suggest the potential ring on the slim finger of the rich lady. He releases it from between his fingers, slides it deftly back into the paper, examines it again by loupe. He takes it again, lets it squirm in his palm, light reflecting sharply, then holds it under the table, where the light is dim, to check for fluorescence. If fluorescence is present, it indicates irradiation, which lowers the value.

Diamond brokers, the go-betweens among the dealers who buy and sell stones, do not own the merchandise. They carry the stones physically and symbolically to prospective customers. They link constituencies, just as Jewish traders in the Moroccan *suq* link the trading towns. They're always on the phone, checking in with the stones' owners, and talking up the stones. A broker must represent the owner's interest as well as the buyer's and must mediate between them to make the sale happen. Brokers depend on the quality and price of the merchandise. Green wrote about the role, "'It's tough to be a broker, you know,' said [a trader]. 'He's carried that stone around all day and got nothing for it.' For a broker, it's all hard legwork. He is the matchmaker, he has to plead and beg—he's got to talk both sides into bending. But what happens? He does $100,000 worth of business and gets $1,000 for himself" (1981:202).

When the broker makes a sale, he or she usually receives a commission of 2 percent of the sale. Although DDC bylaws say the commission should be at least 1 percent, no one—including the dealer—actually knows what the broker makes and what his commission is. Higher rates of commission occur in other countries. The greater volume of smaller stones that change hands more often in Israel accounts for the 1 percent commission on each trade. A dissatisfied and cynical broker in his 50s I talked with barely seemed to eke a living from his work, but he was valued for his honesty.

I shadowed another broker, Henry, who showed a stone to a dealer who peered at it through his loupe. The exchange went as follows: It's very fluorescent, comments the dealer. Henry raises his eyebrows, and the dealer responds, "Come on!" Each of them looks at the stone under the desk, and Henry reluctantly agrees that the stone has fluorescence. The dealer says, 6[000], 6500 . . . to which Henry responds that he has to have 10[000] because the owner paid a lot more. The dealer says that what the owner paid is irrelevant to him. They look at another stone. Henry says they're asking 6500 but the dealer was hoping for 6. Henry says he's gotten an offer for 5750. Henry leaves the stone with him to consider.

Later we go to the DDC and sit with an Israeli buyer who looks over some of Henry's goods. The buyer has a sign on the table specifying the characteristics of the stones he wants. On one of the stones he's showing he tells the buyer that the owner wants 3500 for himself so Henry needs 3570 to realize his 2 percent commission.

Disputes arise over brokering, as they do with every other aspect of the business. An arbitrator told me that he had been asked for advice that day. A buyer and a broker almost had a deal. They had agreed on the price but they were still quibbling about the terms. The broker wanted him to give him half now and half in thirty days; the buyer wanted to finish paying in forty-five days. After a few days of back and forth, the buyer said no to the whole deal. The broker asked: Was that unfair? If the arbitrator thought so, the broker might go for arbitration. The arbitrator thought it was fair since specific terms are important to the deal, and the buyer's offer of a commission was gentlemanly and fair.

One new broker, learning how to sell polished diamonds, was watching his father buy in order to improve his negotiating skills. When he wants to broker a stone, people won't entrust it to him if they don't think he'll do well by it. He must "defend" the stone well and get a good price for it. A poker face is necessary and the steps involved are predictable, but the dealing never goes straight.

Brokers are also a critical link in information exchange in the marketplace. As the lookouts for dealers, they go farther afield than dealers looking for the potential customer. They accumulate and process information about other brokers and dealers and spread the word.

Brokers occupy a dubious status as they flit uncertainly between dealers who own the goods. They are essentially liminal, neither here nor there, but in between. Not only do they not own the goods, they often lack complete information about diamonds in general. Filling an essential role between two sides, these individuals are somewhat stigmatized. Like a marriage broker, they are useful in linking people together, but they are perhaps not completely necessary. A certain invisibility to their role somehow diminishes them.

Additionally, their livelihood is often precarious, as the overall profit margin for the small diamond traders has been reduced. A dealer told me he thought brokers today were barely making a living, as middlemen were being badly squeezed and brokers were the most prominent casualty. Some, however, believe the broker's role is enhanced if he is clever enough to find a niche. The anxiety is intense. Whether a broker makes the sale determines whether he can put food on the table. "The other thing I see much of is jealousy," a former broker, now dealer, wrote me in an e-mail. "While the majority of brokers are struggling to make ends meet, a lucky few are doing well. Some dealers are successful while others struggle. I see this on a daily basis . . . [an] Hasidic friend who recuts stones is wildly successful while his jealous 'friend' (they are not speaking to each other) is struggling to sell his leftover goods."

Some diamond dealers offered a philosophical reason for the lowly status of brokers. A few Antwerp-born New York manufacturers said doing business in Antwerp is much more difficult than in New York because the Antwerp business climate and cultural ethos are bitingly harsh. Furthermore, they said, there was no one lower on the social scale than a broker. A joke from Antwerp went: "A brokerage is a good business because you see a lot of the manufacturers become brokers but very few become manufacturers." This joke underlines the inferior status vis-à-vis manufacturers or dealers who both own and have a lot of knowledge about diamonds. People who go out of business can always resurrect themselves by becoming brokers.

These manufacturers compared the lack of dignity between a *schnorrer* [bum, cheapskate] and a broker. The *schnorrer* has more dignity, they claimed, since there's a certain pride in the profession. To illustrate the point, one of them told a joke about a *schnorrer:* "There were these two men negotiating about a diamond. And they were offering $100,000 for a 5-carat stone. They were apart and they couldn't get together, and there was a commission. And the other guy got stubborn; the seller got stubborn, didn't want to sell. And there was a *schnorrer* there who had gone on his rounds . . . a Jewish version of a beggar. And he's listening to all this, and he says, 'You know what? I'll give you the difference.' This beggar, suddenly? He buys a stone of $100,000? [Seller] says, 'You know, I'm surprised. It's $100,000 cash.' [*Schnorrer* says,] 'That's OK,' and he takes out $100,000, he pays it, right. [Seller] says, 'It seems to me you've got plenty of *chutzpah*: here you go begging and you've got all this money.' [The *schnorrer*] says, 'Well, for me, diamonds is a side business.'" The manufacturer roared with the punch line and rested his hands in his lap. Then, as if to explain his enthusiasm for the joke, he added, "I heard that story in an office in Antwerp where it's properly told."

Then they went back to discussing brokers. "The broker is paid to take the brunt of the business," one of the manufacturers said. "For example, why do we need the broker? Because two dealers sometimes have to compromise. How do they compromise? The broker makes it possible. 'Well, let me try. Let me try to get more.' Know what I mean? For example, this allows both the buyer and seller to take a strong stand one to the other. The broker is the man between. He's trying to make a living, so he has a reason to try to make the business go. That's the reason for the broker. The two bodies can't lose face. What is sacrificed in the process is the broker's dignity. That's the price he pays. He gets a percent or two. And that's the reason he gets his reward. The sale of dignity is 2 percent. Dignity is his stock in trade."

When I related these comments, however, another dealer disagreed strongly and said, "Too harsh!" But the manufacturers noted that the social status of brokers in Antwerp was worse than in New York. They make more money there, however, because of the larger volume. Perhaps the lower status is the trade-off for higher income? The manufacturers recalled with shame their father speaking in a rude tone to a broker in Antwerp as he pushed the broker to get a better price. On the other hand, they concluded, "There are brokers who are quite big movers . . . who are quite able and very skillful and somehow have enough personality to keep the dignity and do it in that way. They have a very firm hand, and they do a lot of business and they're very successful. Sometimes they're quite a bit more successful than the people they deal with. That happens too. But they're not as much fun to talk about." And the laughter resumed.

One observer notes that some large manufacturers can buy stones in huge quantities and operate successful brokerage concerns (Bernstein 1992:147). Some speculate that the role of broker is more significant now than it used to be because of better access to information, but the broker's access and authority depend on the trust and reputation the broker commands. As the diamond business grows more global, how brokers will fare remains uncertain.

◆

Traders in the Moroccan *suq*, Geertz tells us, rarely used "extensive search" to obtain a range of prices for similar items, which were usually new and standard, because most items were not new and standard. An item's worth had to be "intensively" searched to compare it with other roughly similar items, and extensive search was needed to gauge the overall conditions of the market. In the DDC, like in the Moroccan *suq*, diamond traders conduct extensive searches to ascertain the general tenor of the market. They characterize this as getting the feel of the market, the "smell" of current prices not obtainable from price lists. One dealer said people come to the club to get a sense of what the prices are. A person can become intimidated in the offices; you can't compare prices so easily. In the club you can jump up and ask several people from different firms to give you a reading on a price you are using or have received. You get a certain security from hearing the validation from others in the club.

Social scientists call the means by which people move goods, services, or people from one person to another exchange. The universality of exchange is uncontested. Though value is not inherent in things, "Economic exchange creates value" (Appadurai 1986:3). Malinowski's (1922) description of at least eighty kinds of exchange among the Trobrianders and Lévi-Strauss's (1949) examination of wife exchange each spurred vast literature of refinement and refutation. Humans exchange everything considered valuable—whether dinners, trading cards, gifts, political contributions, housework, performances, and baby-sitting—and have diverse ideas about how these exchanges should be made, with what kind of payment, surrounded by what kind of ceremony, at which times of year, and with what kinds of expected returns. Exchanges take place between two people, between groups of people, in sedate surroundings, in raucous marketplaces, and in every other kind of setting imaginable.

The sheer fun of exchange is an important purpose and regulator in marketplaces often overlooked as theorists spin formulas and construct models based on principles of efficiency and rationality. "These sociabilities, the camaraderies of market-places, are universal and are reported from every study," Davis has written (1996:221). "Markets are also parties, meeting-places not only for Supply and Demand but for friends, cousins, lovers, gossips, recruiting officers for factions, collectors on behalf of good causes, supporters of football teams. . . . Maintaining these sociabilities is important, and that is one further reason why people do not cheat, exclude, renege, and create monopolies, as much as they would do if they were solely motivated by profit."

Among friends and colleagues, humor can dispel some of the anxiety that attends the intensity of the trading. Keeping track of one's goods entails difficulties; the diamonds are so small, and they are often lost. Once, when Ben had to give a stone back to Hank, he couldn't find it. He had envelopes in all his pockets and he started to go through each one methodically. He said he was worried about trying the last pocket in case the diamond wasn't there—which reminded him of a joke about the man who can't find his stone and he's looking and looking everywhere. His son asks if he checked his shirt pocket where he always keeps the most important things. The father answers, "No, because I'll kill myself if it's not there."

Then Ben checked his last pocket, and there it was.

The sociable and witty atmosphere in the diamond business is an integral part of the way the trade is conducted and contributes to the pleasure the participants have in continuing economic interaction with one another. That atmosphere was apparent in one exchange between a broker and a Japanese buyer at the club: The buyer slides the heart-shaped diamond off the parcel paper along the folded seam to examine it. He takes it with tweezers and places it on his fingers, then back in tweezers, stretches it out far from him, then brings it closer, then back on his fingers and repeats these motions a few more times, back and forth. "The stone is not very nice, is it?" he finally comments. The broker says, "Yeah, well, everyone has their own idea." The buyer gives the stone back. He says the shape is OK but the make is not good. There is not enough fire, life. They agree that heart shapes are hard to make in a way that retains fire in the stone. The broker says he thought the make was OK but the shape wasn't. They laugh over this.

They enjoy the same contest that makes one alert, alive to nuance, threat, the hidden joke, insult, or even the expression of warmth.

Observers consider it remarkable that the diamond business—global, complex, and involving billions of dollars annually—is an exchange system virtually unencumbered by formal contracts. Breaches that occur, if not settled between the traders, are generally handled by the arbitration system. How can this system rest on such a seemingly flimsy basis?

Social scientists point out that formal contracts alone are rarely sufficient. They are usually supplemented by informal binding mechanisms. Sometimes formal contracts are unnecessary since informal contracts are heavy with moral weight and embody certain principles: to honor commitments, to produce a good product and to stand behind it, and to preserve reputation. Or, put another way, "You can settle any dispute if you keep the lawyers and accountants out of it. They just do not understand the give-and-take needed in business" (Macauley 1963:61). Nonlegal sanctions, such as reputation, the force of conscience, and the power of shame, constrain the actors and direct them into proper behavior (Charny 1990).

Durkheim said, "When men unite in a contract, it is because . . . they need each other. . . . It is a compromise between the rivalry of interests present and their solidarity . . . a position of equilibrium." Contracts can stipulate only so much, however. "If we were linked only by the terms of our contracts, as they are agreed upon, only a precarious solidarity would result" (1964 [1933]:212–14).

Another anthropologist to destroy the barrier between so-called primitive and modern economies, Abner Cohen, described how informal mechanisms of religion, secrecy, and marriage help protect and bind business participants, whether in London or Sierra Leone (1974, 1981). Regarding business elites in London, he noted, "It is evident that millions of pounds worth of business is conducted daily in the City without the use of written documents, mainly verbally in face-to-face conversations or through the telephone. This is said to be technically necessary if business is to flow" (1974:99).

A diamond trader in Amsterdam recalled, "I was once at a notary's office and he asked me what sort of contracts we used. I happened to have a 'contract' with me at the time. It was the inside of a cigarette packet on which I had written an order from a couple of Frenchmen with whom I had gone out to eat the evening before. The only piece of paper I had was an empty cigarette packet, so I tore it open and that was my contract. The notary couldn't understand it" (Lipschitz 1990:137).

Being in contact with one another, continuing relationships over time, and cementing linkages with deals that mutually satisfy is reinforced with values about trust and honesty, concern for personal reputation, and a shared overall interest in maintaining the viability of business.

The Amsterdam dealer whose "contract" was a cigarette packet continued, "But you know, if anyone wants to get out of a deal, if they're not absolutely honest, a contract won't help in the least. You have to be able to trust each other in this trade. You could give someone a package on approval and what you'd get back would be exactly what you'd given. Of course there were crooks. If you gave them stones, you'd get other stones back; only the weight would be the same. But these people would be automatically cold-shouldered" (Lipschitz 1990:137).

The diamond market is famous for its trust. Some trading circles can create strong connections—like a description of the Fulton Fish Market makes clear: "It's equally important for buyers to establish close ties with wholesalers because here everything is based on trust. . . . If a customer stiffed a wholesaler, he would never be sold to again. It's a closed world, a club, like the diamond district, where one's word of honor means something" (Lopate 2001).

If trust is a trademark, what are the safeguards against the risk? The businessmen in London and the Creole civil servants in Sierra Leone confine their trust to a prized few who share similar values, "speak the same language in the same accent, respect the same norms, and are involved in a network of primary relationships that are governed by the same values and the same patterns of symbolic behaviour" (Cohen 1974:99). Ranging in wealth from that of the modest Moroccan trader to the most wealthy London elite, New York diamond traders conduct business with similar principles.

A further factor binding many of the traders together is the shared experience of the Holocaust. Not only have the Jewish traders from Europe been able to continue speaking Yiddish with one another in their work but also the fact of their existence keys each of them to a shared past. The diamond business has attracted Holocaust survivors who can maintain old ties and allegiances. As Helmreich writes about diamond traders who survived World War II, "Frequently, awareness of these shared experiences engendered feelings of mutual trust that became the basis of business relationships" (1992:108).

It was evident to one of the female traders named Joyce that the issue of trust included a gender element. She told me: You have to know who you're dealing with; people trust women but not as much as men. However, she was the only person to tell me that trust had a gender component. When people can trust one another, noncontractual transactions are usually stable. A honed judgment hedges one's bet, and people in the diamond trade are shrewd. One man told me a particular trader was capable but, he added dryly, "has difficulty with the truth." A memo or a contract with an untrustworthy person does not protect; likewise, written provisions are unnecessary for people who are honest. Stories about trust illustrate its significance. An Amsterdam dealer said, "There was a man who once left an enormous package of diamonds with us. . . . We only realized that he had left the stones behind just after he had gone. So we phoned him up to say that we were interested in examining the stones and: 'Before you get a heart attack, you left your stones on our desk!' Yes, and that wasn't the only time that sort of thing happened" (Lipschitz 1990:137).

I came across many stories about trust in the New York diamond business. Ellis Frenkel, a fourth- or fifth-generation diamond dealer—he couldn't remember which—said the amount of trust is "unheard of." Once he didn't realize that he'd pocketed someone's diamond until he got to the airport and he felt it. He called his secretary, asking her to call the person and to tell him not to worry, that he had his 4-carat stone. A memo wouldn't have helped. Every dealer has a story like this.

Many diamond traders minimize their chances of being financially burned by keeping their trading circles small. Though their knowledge of their trading partners is never complete and they have no foolproof way of knowing the financial stability of those with whom they do business, they have fairly certain assurances. Persistent, blemish-free trading over the years provides the best overall proof, especially in retrospect. A jeweler I went to with my cousin provided this perspective: His family firm began in the 1920s or so. He deals with second- and third-generation customers who come to him because the relationships have been established over time, and word of mouth has spread his reputation. Though his prices might look higher than those in auction catalogues, his customers trust that they will generally pay less when they're done than they would have at the auction. He sees his clients at good times, when they want to celebrate, and at bad times, when they need to sell in order to get money.

The framework of trust—based so much on no real certainty—becomes more fragile as the network of trading partners expands. Since trade is essentially this kind of expanding network, pervasive risk and uncertainty always underlie the entire enterprise. A dealer can give a stone to another whose integrity and solvency he trusts, then this dealer shows the stone to another whom he trusts. The first dealer has no control over the people further along the chain of trade, though he or she can try to influence it by questioning who the people are in the series of transactions. If the person down the line, E, does not have the money to pay D, who cannot then pay C, who cannot then pay B and then A, who finally has neither diamond nor payment, A, B, C, and D have learned not to trust E. They will probably not deal with E further, and will pass the information along.

The world perceives diamond traders to be secretive and suspicious, clannish, private, and insular. Cautious, conservative, and careful are less pejorative terms for these same traits that help minimize risks, keep information reliable, and preserve reputations. These characteristics also help people compete, as traders look to guard important information and to outperform the next guy.

When traders branch out to others in a wider distribution, they are admired for their audacity—and watched to see if they are hurt. One dealer told me about Harry, a dealer who travels outside New York a great deal. He does a bold business, traveling with his goods. The dealer I talked with said he would *never* do that! On the other hand, he said, maybe it's not such a bad idea to be bold. Maybe innovation comes from the outside.

People learn from one another's mistakes too, and these mistakes become cautionary tales. I am asked: In that last conversation, did I pick up on the possibility that another dealer was "in trouble"? If the story is true, it is important information for everyone to have. The ripple effect of someone going bad is scary. The dealer supposedly had something on memo to this person, and many are affected. Hearing about someone's financial problems exposes the fragility of the enterprise and the illusion that it's solid when it's faith.

Each trader determines his or her level of comfort. Added to personal trust is the risk a person is willing to take. A young dealer admired a friend's boldness.

"I have a friend who does not buy to look and sell to make a profit," he related. "He only looks to buy if he can recut [the stone] and make a big profit. And he takes bigger risks because he has to recut it to improve it. . . . When you recut something, you take a big chance that something [bad] doesn't happen. [One time] he lost his whole investment because the stone cracked on the wheel. . . . So he has a lot more tension—a different kind of tension."

One's temperament and experience affect perspective. Joe does a lot of business with another dealer who trades with an exquisitely small number of people. Though Joe thinks of himself as cautious, the other trader considers him bold in comparison. Joe told me he wants a larger range of contacts and is willing to take the added risk in exchange. His trading partner values his peace of mind and will not enlarge his circle.

But beyond personal limits for stress, the business is inherently risky. One dealer explained, "First of all, we're playing around with big amounts of money. And the basic tension is whether we're going to be able to make money and whether we're going to avoid losing money. So every aspect of it is very tense." He paused to catch his breath, then resumed his rundown of risks. "For example, let's say we start with looking at a stone. Somebody shows us a stone. We want to offer on this stone. We have to be very careful, thinking about how does this compare to what we have? How does this compare with what we sold? How does it compare to what other people are showing and other people are buying or selling? If you make a mistake about any of these things, you're going to end up losing money—or holding it forever, which is the same thing." Here he stopped briefly, put his elbows on the table and pulled himself closer. "And once we offer on this stone, we're worried first of all, are we going to be able to pay for this stone? I mean, business has to continue on a certain level, in a certain way that we should be able to pay for it. And if we buy it, then the next step is: Are we going to be able to sell it?" Now he was building momentum as he lived the words he was relating. His hands growing more animated, he spoke more rapidly. "And people come and ask us for the stone. So we give it out on memorandum to people. So then next we're worried, are we giving it out too cheap, are we giving it out too expensive that it will never sell? Let's say we give it out for a certain price; will we be able to replace it if it does get sold? And then finally, who are we giving it out to? Let's say they say, 'Yes, it's sold.' Will we get paid? Are these people solid? Will we get paid when they say we're gonna get paid? Then we base ourselves on that, you know, that we're going to get money in thirty days or forty-five days or sixty days. We go ahead and commit ourselves to other things. Will they live up to their commitments? So," he wound up finally, taking a deep breath and letting it out with a faint whistle, "there's a lot of tension involved in every aspect of it, and any one of these things—if they fall through—it will be difficult for us; it will create problems for us." He pushed himself back from the table to look at whether my face registered some understanding of the anxiety he was trying to describe.

A dealer told me a story he'd heard in Antwerp that sums up the constant tension underlying the highs and lows of buying and selling: Someone meets a man who is on his way to market. "Oy," says the man going to market, "I have oxen to sell; I don't know if I can do it. I'm worried; it might not work out," and he goes on whining to himself like this. Later the first man meets the other man coming back from the market and asks him what happened about selling the oxen. The man responds, "Oy, I don't know where I can buy more oxen. . . ."

And the buying and selling cycle continues.

Personal reputation is one of the most important foundations on which trust is built. Lamenting the loss of social connectedness in modern American society, political scientist Robert Putnam noted the advantages of the diamond business. "When economic and political dealing is embedded in dense networks of social interaction, incentives for opportunities and malfeasance are reduced. This is why the diamond trade, with its extreme possibilities for fraud, is concentrated within close-knit ethnic enclaves. Dense social ties facilitate gossip and other valuable ways of cultivating reputation—an essential foundation for trust in a complex society" (2000:21).

Reputation is the most important thing you have, a young trader told me. Reputation is what allows someone to vouch for you, in terms of membership, of course, but also for everything else. Because he thought that people were less trustworthy and that risk was greater than ever, a person's good reputation was vital. Another young trader said: reputation is the main thing. For anyone who cares about his name, this is everything.

If a diamond dealer has a problem with another trader, he tries to determine whether the mistake was an honest one. If the breach was due to something other than an honest mistake, he will not trade with that person and will pass the knowledge to those in his inner circle to preserve the financial security of all involved.

For example, I watched this exchange in which Fred was showing stones to Allan in his booth at street level: Fred ventures, "In confidence, and not naming any names. . . ." Then he suggests that Allan should ask questions of the broker he's considering dealing with. Fred says he's heard some stories. With the first story, Fred gets worried but not overly, but when more stories follow, he becomes more concerned. Allan seems grateful for the tip. Fred tells me later that the broker they were speaking of returned the wrong stone to him one time. A second time he "lost" the stone. Of course, Fred allows, he could be wrong, and that's why he's careful not to mention names.

When a dealer mentioned that another dealer sold his father out for a minuscule difference between his and another offer, his behavior made their firm wary about future dealings with him. Still young, this dealer already had a dubious reputation. Another time a trader told me that a dealer started with nothing, obtained millions, then got wiped out. People were reluctant to deal with him at first after his bankruptcy, but little by little he made it back up and has made himself reputable again. Still another time I watched two men speak in Yiddish at the club. After one left, the other told me he wouldn't deal with him because his son went bankrupt and he couldn't trust that there was enough money there. It is really difficult to know who is OK today, he told me. And if the person with whom you deal is OK, that still doesn't guarantee that the next person will be.

A trader is pegged in different ways. For example, one person may be unflinchingly honest, a tough bargainer but not a risk taker. He has expertise, is aloof from personal squabbles, is repeatedly voted as arbitrator. Another trader's father and uncles were well respected, but this young man and his brother have suffered losses. Talk circulates: How solid are they? Are they overextended on credit? Are they involved as creditors in the bankruptcy case that just broke open? If so, that will directly influence their ability to trade.

I observe two traders in an office. Sam has come to this office to pick up a stone he loaned to Jack. Sam and his family dealt with Jack for years, but about twenty years ago, Jack's family business went bankrupt and the brothers separated. Jack jokes about how he can't find the stone. Maybe it's downstairs in the vault, he suggests. He asks his secretary if she put it someplace. Sam and the secretary exchange weak smiles. The secretary finally finds it. They exchange news of each other's families. Sam looks through his memo book for the memo that has recorded that Jack had the stone; finding it, he tears it out and tosses it to Jack. Jack asks Sam if he has a 4- or 5-carat stone, emerald cut, around a G color. Sam thinks of two that more or less meet that description. They talk about possible prices for a moment. Sam says he has to check. Sam also has a necklace that maybe Jack would be interested in. Jack is. After we go, Sam explains to me that he'll have to check with his partners first about whether Jack can be entrusted with the necklace, which is a major piece. He's thinking of the recent bankruptcy and recalls it's also been a little hard to get things back from Jack.

Family history keys people to a historical store of pertinent information, but everyone knows sons and daughters of trustworthy people who were lax or dishonest. Some of the men in an office were looking over the poor printing quality of the accounting books they had just received. One of the men proclaimed: This is terrible! Another agreed: We really need a new person to do these; these are no good! The older man in the firm muttered that he used to do business with the grandfather. Then deal with the grandfather, protested his son, hoping his father would laugh.

Another time I heard a father and son discuss a firm that hadn't paid for a stone they sold them. Chaim and his father had dealt with the father and uncle for fifty-seven years. The father and uncle had been part of a big and reliable firm, but they had died more than twenty years ago, and the sons of one of the men had been in charge since. Chaim and his father had watched the firm go downhill for years, and now they were worried that they had been burned. Thinking that continuity in a formerly trustworthy family will prevail can trap you. It is always tempting to transfer the knowledge one has of one generation to the next, but it can be dangerous.

A young broker, the fairly inexperienced son of someone they knew well who died, gave their stone to a third party, who shipped it and it was lost. The shipper wouldn't take responsibility for it, and the young broker was not a member of the DDC so arbitration was probably not the answer. After the man's father died, the business had gone bankrupt. One should not presume the financial soundness and stability of a firm just because those traits were evident in the prior generation.

Each case has to be evaluated on its own merits. A firm with several partners has the advantage of containing people with whom one can test one's judgment.

A father and son discuss whether to give a stone to someone they've never dealt with. The father has known the person for thirty to forty years but is hesitant. They gave a stone to another person last year and couldn't get it back for about a year. When they finally got the stone back from him, they decided never to deal with him again. Something about that guy had made them worry. Sure enough, the guy has now been "busted" because he hasn't paid a number of people. The DDC office received a number of complaints about him, and now the whole trouble has been exposed. These men feel lucky to have gotten their diamond back.

A virtue of the marketplace is the number of people from whom one can gather information. With the information, one can attempt to sort fact from artful elaboration, evaluate the stories, and emerge with a fairly coherent account. Despite and because of the "noise" in the DDC, like in the *suq*, the collective continues to judge the worthiness of individuals as trading partners. I was at a table in the DDC when a few people were talking about someone they refused to name:

Now that it is March and the Christmas season is behind us, they are saying, things have "settled": it begins to become apparent who is and who isn't "in trouble," whether a person can't pay his or her debts. This person is considered nasty in addition to possibly being "in trouble." "He's hurt people through the years," comments one. Another dealer says she doesn't trust him. A third person learns that the man they're talking about is the cousin of someone he's dealt with who was difficult. The relationship helps explain the similar temperaments, they agree. On the other hand, the son-in-law, they agree, is a good guy. But the father didn't pay her father for a stone years ago! The case went to arbitration, and he was supposed to pay a certain amount every month, which he finally did.

The traders continued to construct the history, virtually all of which was damning. Such "reputation bonds" are crucial, and no one wants to sacrifice his or her good name. The moral power of reputation helps ensure that a trader will behave ethically. Like prisoners who play a game together knowing they will most likely have to continue to play together for an extended period (the "tit for tat" strategy described by Sugden 1986), the incentive for Moroccan *suq* traders, Fulton Market fishmongers, or New York diamond merchants to behave honestly is rooted partially in the knowledge that their behaviors will accumulate in a textured fabric of reputation history. An editorial in the *New York Times* lauded this tradition of trust that means "diamond merchants are smart enough to understand how much strict adherence to shared values—basic moral sense—enriches everyone" (Starr 1984).

Another kind of reputation concerns a person's religiosity. Many in the diamond business respect those, especially Hasidim, who appear strictly religious. On the other hand, religious piety does not substitute for ethical business practice; in fact, expressions of piety without ethical behavior garner considerable contempt. Traders are too savvy to let superficial emblems of religiosity blind them to dubious marketplace behaviors. Practical survivors, traders find that their Judaism roots them in skeptical this-worldliness.

We observe a group of Hasidim standing nearby in the club, talking and gesticulating with excitement. I wonder if they are talking about the "trouble." But no: the man who is in trouble is an upstanding Hasidic man. But the trader telling me this is wary. She predicts there will be the usual stories (she's heard too often): he's an honest man, he'll make good, his wife was crying in *shul*, and he has eleven kids. The part about eleven kids is always trotted out as if it's the end of the discussion, she adds sardonically. And big deal that he has eleven kids. She heard another Hasidic man say, "Yeah, well, I have nine kids and he didn't pay me either."

Thus is reputation multifaceted and complexly determined. Behavior is unavoidable evidence. The information is individually interpreted and is spun by the group into other incarnations. The wise store pertinent information about reputation against which to test new experiences that in turn lead to new judgments.

Given the need to update relevant information about trading partners and the marketplace, information must be sifted from many sources. Traders must be able to procure, use, remember, and adjust information about traders, firms, prices of different kinds of diamonds, and criteria that affect the evaluation of diamonds in hand. They must have a workable store of information about the international markets in general and the changing consumer trends in fashion, design, and luxury goods. They need knowledge about the production of the commodity as well as knowledge about the external market, the consumers, and the world. The bourse grapevine is an effective and efficient source of reliable information for the diamond trader.

◆

Despite the importance of the bourse, the DDC was losing some of its vibrant and central role in the business, and many participants were lamenting the loss.

Usually the first thing I would hear when I hadn't been to the club for a while was that I wouldn't recognize it because it was so empty. Older traders expressed sadness and resignation. They seemed surprised that the bustling activity of so many decades seemed to be over. The club is dead until about 11, 11:30, noon, one told me. It picks up for a few hours, but the number of people there is still a fraction of the numbers who used to come. People used to rush to the club by 8 A.M. to claim the best spots by the windows with the best light. No more. Hints of this trend were readily apparent in the 1980s, but in the late 1990s many traders noted the lackluster activity in the DDC, and the quiet of the club seemed particularly ominous. A past president of the DDC noted that the decline in bourse business was a worldwide but dangerous phenomenon, adding, "Today, when the industry speaks, we have a lot of clout. But if the bourses decline, we will lose much of that power" ("DDC Efforts Seek Higher Profits for Members," 1995, 32:82).

Informants cited four related reasons for this change. One, as the diamond business has grown more globally controlled and interlocked, increasing amounts of the business are concentrated in fewer and larger hands. Two, this process has led to progressive squeezing of the middleman, the small trader, and the tiny family businesses, so that increasing numbers of these individuals and firms have gone out of business. Three, as larger firms have grown even larger, they have tended to conduct their business in one another's offices. When these firms host foreign buyers, they monopolize and seclude them on their own territory. As they hoard the business, there is less business for the smaller traders, and the spiral of decreased DDC activity continues. And four, the biggest player, De Beers, hastens the process worldwide by dealing with fewer firms, increasing the pressure to commodify the diamond and to concentrate business further. De Beers's going private and competing head to head with them is likely to exacerbate the situation.

Brokers and dealers protested these developments with their stories, complaining that bigger dealers shouldn't siphon off the DDC trade. A broker who often sponsors foreign buyers in the club described a serious breach. His buyer was sitting at his table with him when a prospective seller slipped the buyer a card and said, Come to my office, where I can give you a better price; don't pay his commission. The sponsor, who would receive a commission to compensate for the expense of bringing the buyer here, was outraged. In fact, he took the matter to arbitration. People aren't supposed to steal your buyer, he told me. DDC members are deprived of the opportunity to sell their goods when the buyer goes to private offices. In this case, some sixty people sold to this buyer in the club. Obviously they would not have made those sales had the buyer been sequestered in someone's office. As it turned out, many people in the club were behind the broker who sponsored the buyer, and he won the arbitration. Since buyer stealing was becoming quite common, he believes this arbitration helped put some brakes—if only temporarily—on the practice.

The ruling regime of the DDC was given partial blame for the decreasing business in the club even though the board was attempting to infuse new life into the club. One prominent member was angry about the officers' behavior.

Basically, he told me, the present administration doesn't want the club to have buyers, and they'd rather see individual offices have the buyers. Even though other bourses don't operate like this and it wasn't always this way in New York, the trend of doing business away from the club is unmistakable. He pointed his finger forcefully, for emphasis. Your customer, whom you bring in, should have the right to buy from you, and then you should bring that person to the club, where everyone else can offer their wares to him too. You should get the commission, no matter what, from everyone who makes sales from your buyer, including if they occur after he leaves, because you're the one who was responsible for bringing him here and setting everything up. Now he spoke slower, for he was getting to the most important point. People are not allowed to steal your buyer, period. There is a reason for this: it's better for everyone in the long run to have this active marketplace rather than have the trades sequestered among a few people in outside offices. It generates more business for everyone. This is why the club is deserted. It all started changing in the 1980s.

From his point of view, the DDC administration started favoring some private offices over the well-being of the entire group at the club. Another dealer charged that a recent director had brought glamour and prestige to the club and to the diamond business in general. But as he brought more people to New York, he took them to private offices, including his own, and ended up skirting the club, the very institution he was supposed to represent and enhance.

One person who was not alarmed at the trend put it in context with his basic philosophy. He said he believes that making a living is always challenging. Among some men who were arguing about business being diverted from the club to the offices, he alone said he did not think it was a DDC election issue. "Business will happen wherever it happens," he said, shrugging. Others disagreed. Meanwhile, a dealer in his 70s said he believed that this issue was nothing new:

Yes, the administration is always out for itself, he said, his partners nodding vigorously. It has always been thus. Little is ever done for the greater good, as far as their experience tells them. It's probably one reason they rarely go to the club, preferring to manufacture for their tried-and-true customers here and abroad.

The philosophy that business is always hard is historically true. As supply and demand have waxed and waned and as new problems in the industry have arisen, there has always been room for the entrepreneur to recognize opportunity, to adapt to the changing environment, and to create niches. When the first diamond traders came to New York from Europe in the 1930s, they had to build the business. Opportunity was everywhere, but it was difficult. Few people helped them. They trained cutters, and the cutters left and started their own firms. New methods of manufacture provided threats and challenges that were exploited by the right people. Prior to the development of the labs, inexact standards for evaluating and comparing stones permitted some easy profits because consumers knew little about diamonds; abuse led to FTC crackdown, consumer distrust, and further difficulties. Boom and bust cycles in the national and worldwide economy have created millionaires and have wiped people out. One has to be both cautious and bold, trust old friends, risk some new ones, adjust and evolve. Anyone who makes reference to some golden age when profit was easy and downside was minimal has allowed the mellow glow of nostalgia to cloud memory. Wistful, wishful thinking is definitely not adaptive.

◆

A promising development in the late 1990s created a flurry of excitement—if only short-lived—at the DDC. The DDC administration sponsored several "Market Weeks" to help bring business into the club. Nestlebaum reported a "surge of business" from fifty-seven new firms brought into the club during one of these weeks (1996). "For the first time in years, the Club was packed. Some people who hadn't been here in years came up just to see it with their own eyes," she reported that one of the organizers said. More than 650 stones changed hands, according to the official weighers, and more deals resulted later. Trumpeting the success, the DDC planned to repeat the event several times a year (Shor 1996b).

Another report was as follows: "When the Diamond Dealers Club of New York held its first Market Week at the end of last year, it brought more than business. It nearly brought tears to the eyes of the Club's old-timers and long-timers, who felt as if they were being magically transported to the good old days. They saw familiar faces that hadn't been seen on the trading floor in years. They saw crowds; they heard the delicious din of deals being made at row after row of packed tables. . . . In those few days, the Club did an estimated five times its normal business and untold numbers of new business relationships were forged. But most importantly, the Club had broken through the perception barrier that was leading major New York buyers to foreign centers before even bothering to try their own backyard. Finally, they saw that the goods were in New York, available, well priced, and just a stroll or an elevator ride away" ("The DDC: Gateway to the Diamond Heartland," 1997:4).

Organizers of the event hoped that Market Week would transform the DDC into a lively hub, and the event was repeated several times. Some had ideas about expanding the scope of it. A female dealer, wary that the event might not be repeated, wondered if it did any good—but allowed that if some people met other people, that was good. A man in his late 50s was enthusiastic, saying that it boosted morale and benefited everyone, especially brokers. A man in his 80s also thought the event had gone well and hoped it would be repeated because people profited and many new contacts were made. How good it was to witness old, established firms come to the club, see goods, meet brokers, and conduct their excellent business once again in the DDC, he said. How symbolically important it was to counteract the belief that business could be done only in the offices. Maybe, maybe. . . .

◆

A busy marketplace, full of cacophonous voices of dissension, joke telling, and deal making, pulls the trader into the fray. The emotions are intense, juxtaposed with what feels like interminable waiting and fairly constant anxiety, many traders told me. Mixed in with the anxiety and the difficulties are the unmistakable pleasures: the ecstatic highs, the lively sparring, the comfort of simply connecting with others, the jokes. I was standing with a group of traders in the club when Sammy joined us.

Jack begs Sammy, an Iranian Jew, to tell a joke because he's the best. Sammy doesn't want to, but as the others talk, an almost imperceptible change begins to come over Sammy's plastic face, and he starts to relate a story about how a very religious man is standing outside a delicatessen, drooling over the wonderful *traife* [non-kosher] things that are inside. He sees the plump shrimp, the different kinds of wonderful hams, the lobster, the thick Canadian bacon, and delicate Italian prosciutto, and each looks succulent and inviting; as he looks and gazes, little droplets of spittle start to form at the corners of his mouth. Sammy plays it out, the planes of his face lengthening and expanding as he tilts his head up to continue. Finally, the man cannot stand it any longer and he goes inside and orders a large platter of everything that he wanted: the hams, the shrimp, the lobster, the bacon. And it is set before him like a banquet, and he eats and eats and eats. Finally, he lifts his face to take a breath and he sees the rabbi—from his *shul*, no less—standing outside the delicatessen, looking in through the window. The man gasps, wipes his mouth with his napkin and rushes out. "Rabbi," he says, "Did you see me in there?" The rabbi nods sadly and slowly. "How long have you been standing there?" "A half hour." "A half hour! You saw me the whole time?" "Yes." "Oh, thank God! I ate under rabbinic supervision!"

Uproarious laughter met this story. Far from diamonds, it nevertheless cut the tension neatly. The joke masterfully plays on the anxieties of conforming to burdensome rules, the difficulties of being moral, the strain of being seen, and the seduction of temptation. These typically Jewish concerns are inherent to diamond dealing, too, and the joke skewers them efficiently and ridiculously.

The undercurrent is constant: Did I sell too low? Should I have waited for a better price? Will he come through with payment? Can he be trusted with the stone? Should I offer on the pear-shaped though I might not be able to sell it? Did I seem too easy; should I have been tougher? The momentary elation of having bought or sold a stone is followed immediately by the second-guessing: It felt so right! Did I do right? Another joke brings out the sadistic character of this competitive anxiety.

"This brings to mind another colorful character by the name of Finkelstein," a manufacturer related, attempting to stifle the anticipated pleasure of the joke welling up in him. "Finkelstein was very tough, had a very tough attitude. He was pitiless. Once sitting in his office, one of his customers, a dealer who sold rough goods to manufacturers, came in to buy, and he complained about what Finkelstein had sold him the last time. The customer went through this whole litany. He tried to explain what he had gone through, and through the whole story, Finkelstein listened very sympathetically. At the end of all this, Finkelstein told the customer, 'You know what, come and tell this to my partners.' So they come into this inner room where they all sat around, and the customer repeated the whole thing with all the complaints—and when he was all done, Finkelstein turned to his partners and said to them, 'You see, you couldn't make *any* more!'"

Here the customer was *ruined* by the deal (or complaining to that effect), and Finkelstein's response is to brag that he made the best profit he could. What appears at first to be patient listening is turned around to mock the man who complains. Instead of getting a sympathetic ear from Finkelstein and his partners, he ends up justifying Finkelstein's prowess. The manufacturer who told this story thought it captured the cutthroat competition. What relief the joke provided! It also underscored dual pressures that a dealer might often face: staying alive financially through his wits and having to prove himself to closely scrutinizing partners, who are almost always related to him and exquisitely attuned to his flaws.

A young diamond dealer told me how it had been for him recently. "We bought something that we thought was good. We overpaid. It happens every day. It happens the opposite also. You bought it; you thought you'd do one thing with it. . . . I once bought a 2¼ stone, an imperfect stone, and by itself it's really not good to have an off-size, a 2¼, because people usually ask you for a 2-carat. And between a 2-carat and a 2¼-carat it's a 10 percent difference in dollars. So I held it for a while—I'd say a half a year, maybe even a year. And then someone showed me a similar stone that matched. I offered on the stone. It was a manufacturer and it didn't come out right for him, so he had to think about it. And the price that I offered was 10 percent more than I paid for the stone that I held for a year. So I was really paying a full price for it." He let the impact of this last statement sink in on me as well as his partners, listening and nodding throughout the story. Then: "I sold the pair of stones within the day for much more! And you know, you think it's a *strop* [bad deal]; you think you paid too much; I'm going to hold it. And I find a match to it, and I [snaps fingers] sold it immediately! I could've been stuck with two stones, double my problems, and one would've been 10 percent more."

Manufacturing introduces its own set of serious and constant concerns. As the manufacturer plots how to cut the stone, he or she realizes the gamble that is entailed. A wife remembered that her husband had sleepless nights worrying over various stones, whether cleaving it this way would work or destroy the stone, whether removing an inclusion would measurably improve the look of the stone, result in too much wastage, and so on. The countless decisions are not always easy to live with. A manufacturer in Antwerp put it this way:

"We live with calamity and must be philosophical about losses. With the most expert knowledge we indicate a cleavage line. Then a stone falls apart along other lines, costing us a small fortune. . . . There are other times when we live with disaster. . . . Occasionally gas under tremendous pressure is trapped in a pocket in a diamond. When these pockets are struck, a valuable stone may explode into dust. Several months ago I had a 5½-carat diamond worth $11,000. There was a minute imperfection at the edge of one of the facets. A purchaser wanted it polished away. The stone had barely touched the cutting wheel when it shattered into a thousand pieces. Yet we don't have bad luck all the time. Several years ago I bought a 35-carat brown diamond. It looked virtually worthless, but I hoped some small fragments might be of gem quality. To my amazement—there was no way to know this in advance—all the brown discoloration was concentrated in that piece. The remainder was flawless blue-white of highest quality. It polished out into a beautiful 12½-carat stone" (Ratliff 1959:227–28).

Even though those who buy and sell are spared the anxiety about the freak accident that destroys a stone, they too face constant worry. Then there is the tension of keeping track of the little gems, the constant fear of misplacing or losing stones. One manufacturer cavalierly told me about this aspect of the business.

Diamonds can get lost or break off or loosen suddenly from the wheel when being cut, he said. Crazy things happen. Once Sidney heard a strange sound by the safe and thought nothing of it. It seems a stone had hurled itself across the room into the other room, and no one knew where it had gone. Later he was trying to close the window and couldn't do it. When he inspected the window more carefully, he found a diamond stuck in it. It had been flung from across the other room and lodged there during polishing. Sidney could have been killed if he had been sitting in his regular chair. Another time he heard a noise in the box behind him. Later the men in the shop were hunting everywhere for a lost diamond, a 2½-carat emerald cut worth about $20–25,000. When he realized what they were hunting for, he suggested looking in the box, and there it was, way at the bottom, all by itself.

One of the dealers—always appearing serene and even-tempered—told me how much worry he always felt in the business. There is much pressure and it is always emotional, he said. When his father died a few years ago, he felt more of the burden than he had in all the years in which he had been part of the family firm. Now that his own son has come into the business, he feels that he can begin to relax as the son begins to take some of the responsibility from him. So the other day he was so delighted to hear his son say that he was nervous about a stone. He'd said he was worried he'd paid too much for the stone, to which his father replied, "I can't tell you how glad I am to hear that," which made everyone who overheard the comment laugh.

Counterpoised to the tension is the pleasure of the social aspect, a formidable but fun attraction. This element of the business is a sensuous cushion that buoys and supports—despite or perhaps because of the varied and sometimes disagreeable personality traits of people in the club and the continual apprehension of the business pressure. Poking fun is a way to compete. A Japanese buyer, Takeda, spoke Hebrew and inserted Yiddish in his English to be silly and to catch people off-balance—and maybe to better communicate with the Hasidim. He told me that everything in the business is *bubbemeise* [nonsense] and shouldn't be taken seriously. The strain of dealing with or being in the same company with people who complain or are annoying, obnoxious, boastful, unscrupulous, stupid, dishonest, petty, or slow is one thing. But jokes and camaraderie spring from these experiences and people.

A dealer told me of his enjoyment in the business. First, he said, he derives great pleasure from the gems themselves. He also loves the camaraderie of being with the people day in, day out. He feels the safety net of being nestled with his family and with Jews. As he's become more religiously observant as an adult, he increasingly enjoys the security of being among Jews. He tells about an incident: Sitting in his office with his father and his uncle, his father suddenly farted loudly, and he and his father started to laugh. His uncle across the bench was hard of hearing and wanted to know what provoked the laughter. That made father and son laugh louder. When the uncle lost his temper, the father yelled into his ear and then they all laughed together. That kind of thing is terrific, he sums up. You can't beat it.

Even though another dealer said he enjoys the business less now than he used to because of the current tough economic climate, he said he still loves the social aspects. What he loved so much back then was the excitement, the not knowing what was going to happen next, the cutting, the selling, the people who by and large were mostly nice. Even today he enjoys the game of the whole thing, the *schmoozing*—that's the main thing—and the reason that's enjoyable is that the people are worthwhile and good. That hadn't really changed over the years.

One relative who briefly tried the business also recalled that he had enjoyed the game. Though he had not been eager to wear a tie to work, he reluctantly bought a loud tie. At work he was talking with a dealer who said he liked the tie, and that began an exchange between them. The tie was a rapport-maker, a conversation-starter. What made him seem so perfect for the business was his almost innate ability to probe and to maximize the game-playing aspect of it.

This social aspect of business gets mixed up with game playing and negotiation. It is negotiation made palatable, sweeter, more interesting, and more textured. I wandered over to a table where I recognized a broker. Stewie and another broker, Shmuel, were arguing over a stone. Stewie had done a cachet on it the day before, so the deal was pending. Stewie's father sat across the table from Stewie. Shmuel was standing up, pleading. Stewie asked his father what he thought. Stewie didn't think the stone was worth more. Shmuel looked longingly at the father for help but the father said, "I only write the checks; he's the boss." Shmuel wanted Stewie to offer more so he could make a decent commission, but Stewie said Shmuel should demand an adequate commission from the seller instead. Stewie remained firm and pleasant, clearly enjoying the routine. (He told me later, "I've had excellent teachers.") Then he offered to pay less but to pay it quickly, to which his father hissed to him, "We don't have that much money right now." As the prior cachet was torn up, they finally decided that they would pay in thirty-five instead of forty-five days. The *"shlepp,"* or waiting time for payment, would be shorter.

Another time Stewie "blamed" his father for the selling price he had to stick to that was higher than the broker wanted to pay. Stewie "agreed" with the buyer that the price was too high but said they had another buyer in Israel. Doubting, the broker called Stewie's father on the phone and got the same answer. The broker called Stewie's father "cheap" and Stewie, enjoying the game, countered with "careful, shall we say." Stewie then called his father "stubborn" while the broker repeated "cheap." Scapegoating his father allowed Stewie to pretend to commiserate with the other broker, all the while sticking with the price and making the other broker take it or leave it. He could blame his father and get off the hook—or make the deal—with this kind of alliance. Sometimes he uses this approach just to buy time for himself.

At the DDC I listened to a dealer arguing with a broker on the phone. They disagreed about the price the dealer wanted the broker to get for him. The dealer repeated the price he wanted. But then he realized that the two of them differed by only about 1–2 percent, so in the end, the dealer gave in and consented to the broker's price. "I'm a softie," he told me, hanging up the phone. Then he added that for such a small difference it is important not to sacrifice the sale. He knows people who let a 1 percent difference between them derail a sale, and it's not worth it. The point is to buy and to sell, so bending a little is important to maintain the process—not to mention the goodwill.

The dealer understood that he was ensuring a future relationship by agreeing to the less-than-ideal compromise.

Another broker told me she gets pleasure from dealing with and pleasing her customers. She was a good listener and found she was good at the relationship part of the business.

Complaints, on the other hand, occupy a special niche of pleasure about shared difficulties that bind people together. A culture of complaining is pervasive and ordinary. People enjoy it and indulge it. Jews are known for it. A joke—unrelated to diamonds—that I heard in the club went like this: There are three women sitting around. One says "Ach," a hand holding her shaking head. The second says, "Oy," also holding her head and groaning. The third says, "Oy, oy, oy," with even more emphasis. Finally, one of them says, "OK, enough about the children."

Like one's children the diamond business provides endless reasons for complaints and discontent. I asked a dealer why people constantly griped about the shortage of goods. Beyond the real shortage of goods, he said— and it is real—no one is allowed to express satisfaction about his situation. It's a game. Someone gives you a price, and you say, "Ach, I'm getting a headache; you're making me sick." And they go back and forth. It's nothing serious; it's like a duel with some light thrusting, feeling each other out, testing—then maybe a serious jab.

This is performance. Everyone is "on" at the club because it is public. To some, insults are an art form. "Well, my uncle Marcus was a broker in Amsterdam," one of the manufacturers began. "And he liked to tell stories. He was a successful broker, and one day an out-of-town buyer came into the diamond bourse in Amsterdam, and he was a rather obnoxious type, very aggressive. And usually when an outside buyer came to the club, there were all kinds of dealers and brokers milling around because there was a new face in town. And this fellow was rather rough on everyone. At one point a broker sat down to show him diamonds, and he said he wants big stones. He came to Marcus. 'Do you have any *grobbe schteine*? Big stones. 'Big stones' in Yiddish or German is *'grob.'* 'Grob' has a double meaning which means large or gross, common, vulgar. So Marcus says, 'Yes, I have large stones—as gross as you, as *grob* as you.' It brought the house down."

Even when deal making is private, people know whether you are behaving correctly, suspect whether you are succeeding, doubt whether you can be trusted. The public aspect of diamond dealing adds a special fervor to talk and joking. And it is an added pressure. In a profound way, one's life is an open book. The following Antwerp story plays on the discomfort of having everyone "know your business."

"This is an absolute favorite," the dealer told me. "A Yiddish one with the flavor of the *shtetl* [East European Jewish village] in it. There was a story about a diamond man, Lubelsky, an Antwerp diamond man. He lived in a period when people were more colorful than today, though there still are colorful figures today. He went regularly to the site in London from Antwerp, and he told a friend of his that he was taking his wife with him to London. So his friend said, 'I don't understand. I was under the impression that you don't get along too well with your wife.' 'Well,' he said, 'I'd rather take her with me than kiss her goodbye.'"

The spectacle of performance takes on heightened intensity in the marketplace of the DDC, where so much of what occurs is essentially on stage. Hearing stories about how deals go sour between old friends can be enjoyed as performance as well. A manufacturer was delighted to tell me this one: "Two of the tycoons in Antwerp—the rough tycoons, the really big ones—they're close friends, but they do each other, as a family will do. One day—I don't want to mention any names; I know them both very well—One day one of them—the two of them—are supposed to go into a deal together, and one of them finds out that the other one went into the deal by himself—and cut him out! Something that's rather frowned upon—but done. Right. So the one who is aggrieved confronts the other one and says, 'We're old friends; how come you did this to me?' *'Because I couldn't resist.'* True story. Couldn't be angry at them. They're old friends. There was big money involved."

The theatrical aspects of confrontation, snide aside, protest, withdrawal, comic relief, and time-out for politics, health, and family are played out in exaggerated, almost caricatured, form for the benefit both of the participants and the audience of those standing nearby.

One of the brokers, Harry, calls out to me aggressively as I walked toward the table, "Am I going to be in your acknowledgments?" he booms. I approach with some trepidation. But I realize I don't have to answer him, that it's all for show. I nod, just a little apprehensive. But he's interested in the business at hand. He has brought a buyer from East Asia, a Mr. Joseph. People cluster around him to give Harry or the buyer various goods to see. Harry yells and carries on; he talks about his anxiety and how he has had sleepless nights during the last few nights. It's unclear to me at first whether he's referring to the latest spate of terrorist attacks in Israel or to the terrible prices that potential customers are offering. Now he screams, jumping up and down, arms flailing, "Give me $4500, I'll give you a check upstairs today right away! HEL-lo! HEL-lo! HEL-lo!" One dealer nails him with eye contact and tells him to "go sit down and relax." But Harry is irrepressible. He argues with a young Hasid, and they begin to shout at each other. The Hasid believes he has spotted a flaw in a stone that justifies lowering the price, and Harry hotly denies it. "Go check it out!" he rails. The Hasid consults a gemologist in the club and is proven correct. Harry ignores him. More parcel papers are passed back and forth wordlessly between Mr. Joseph and the other men milling about. Harry looks increasingly exasperated. One person offers a price, and Harry starts reciting the *Kaddish* [prayer for the deceased] in sarcastic response. A Hasid looks heavenward, rolling his eyes. Someone mentions *HaShem* [the name, referring to God] and Harry blurts outrageously, "Who cares what *HaShem* thinks!" Low prices are offered as Harry repeats again and again his desired price and how he'll get the check right away if only someone will offer something decent. He dramatically pleads with his buyer, "Please don't be mad at me, Mr. Joseph. I let that seller get away, but I couldn't help it, Mr. Joseph. I couldn't do it with that one, but I'll make the sale for you, I'm sure."

When I spoke to my cousin about Harry's behavior, he said that he's a man he trusts, full of a lot of bluster and explosives but basically fine. His behavior was just normal stuff.

Beyond the tactic of feigning acute displeasure at a bad price is the taboo against acknowledging good fortune. A dealer explained that it's tongue-in-cheek. One hisses *keine h'ora* to rebut the evil eye if complimented on a beautiful child, for example. The standard response to "How's business?" is to complain, to shrug as if one is bearing up but struggling, or to look sympathetically upon the other as if sharing in the profound burden of the business. Business can be terrible—or worse than terrible. As one of the manufacturers once quipped to me, "Everyone in the diamond business always complains. My father always used to say everyone in the diamond business lives fantastically well from their losses."

A dealer told me, "A superstition that comes to mind is that people will never admit business is good—no matter what. *Keine h'ora* [against the evil eye] if they say something good. 'It's OK,' they might say, about the business. No, others object. Business is terrible. It's a struggle. It's considered poor form to say business is good. It's a violation of good manners. It's actually derogatory. People get insulted if you say it's good. People think you're making fun of them, you're saying it's good: What do you mean by that?

"You know the old story," he continued. "A man is going on a trip. A friend says, 'Where are you going?' 'I'm going to Lemburg,' he says. Says, 'You're telling me you're going to Lemburg so I should think you're going to Warsaw. But I happen to know you're going to Lemburg, so why are you telling me this story?'"

Basically, you don't tell your competitors how things are going or with whom you're dealing. Trust is delicate, suspicion and competition ever present, the wariness never off duty. But as serious as it all is, it's also not taken seriously, a dealer reminded me.

Complaining continues a long and hallowed tradition in Jewish culture. Barbara Myerhoff's beautiful book, *Number Our Days* (1979), described the lives of elderly Jews in a senior citizens center in California. She described the intense infighting among the center members and pointed out that sharp verbal disagreements were not only publicly displayed but also cherished—further, they were a social adhesive. "We fight to keep warm," one of the center members explained.

Myerhoff recognized the conflict as a definitional ceremony. "Eventually, I recognized in Center fights a particular cultural style, and I could identify a working equilibrium underlying people's raucous, passionate lungings and lamentations. Quarrels were contained. Deviance was punished. There were regulations, common beliefs, rewards, and punishments. . . . Here was a community, then, sewn together by internal conflict, whose members were building and conserving their connections using grievance and dissension" (1979:187).

The performative features and the social issues evident in the diamond marketplace of the DDC are definitional ceremonies that provide pleasure and release. The work is filled with tension, the stakes are high, the goods are short—it is difficult to buy and sell. The information in the bazaar is confusing, full of noise, distracting, and important to decipher accurately. While competition is fierce and the risk of failure ever present, the social milieu of joking, complaining, *kibitzing*, and *schmoozing* that embellishes the dance of deal making reminds all the participants that the group is necessary for their continued survival and that pleasure is to be had in staying together.

◆ ◆ ◆

[6]

With Loupe and Internet

The diamond business is a fundamentally traditional trade, steeped in rituals that seem to ignore the modern world—but the fact is that the ancient traditions endure alongside insistent modernization.

Two innovations in particular have had an enormous impact on the diamond business on 47th Street and worldwide. The GIA's grading reports and the Rapaport price lists together have created a sweeping change that diamond traders emphasize has irrevocably altered the diamond business. These developments have drastically expanded the availability of information along the industry pipeline from producer to consumer.

In 1981 David Koskoff described the industry as follows: "The wholesale diamond market is very much an informed market in which buyers understand the qualities of the goods. . . . The retail diamond market, on the other hand, is what economists call an 'uninformed market,' one in which the average buyer has an incomplete idea of the nature of the goods he is buying and judges their qualities not by any really understood standards, but on the basis of advertising, reputation, eye- and ear-catching features. The uninformed buyer, the typical retail diamond customer, is at the mercy of his retailer" (1981:279).

Such is hardly the case in the new millennium. For one thing, the 1981 diamond market crash was a watershed for many. Over the next two decades, diamonds were steadily demystified. Lubin called the gemological certificates and diamond price lists among the most "sweeping changes to touch the diamond trade in recent years" and noted that "although certificates seem to have stabilized the precious stone industry in some ways, they have brought a whole new realm of difficulties to trading" (1982:58). Most in the industry would agree. Since their introduction, the lists and certificates have been regarded as possible replacements for a dealer's skill and knowledge.

The GIA grading reports and the Rapaport price reports began to level the trading field between wholesalers and consumers, provoking a trade writer to note in 1993, "The GIA diamond grading terms and standards have been so ingrained in common use throughout the American trade that consumers now ask their jewelers for a specific grade of diamond, e.g.— 'F VS1'" (Shor 1993:228). The industry has been trying to adjust to the implications of these innovations ever since.

◆

The GIA was founded in 1931 and provides training in gemology and research in gemological analysis. In the 1950s, GIA Chairman Richard Liddicoat developed what the GIA terms "the first internationally accepted diamond grading system," which "provides unbiased opinions of the quality of polished diamonds by applying uniform criteria to their grading."

The impact on the trade has been huge. An elderly dealer told me that the old system lacked the "alphabet" that is used today and relied instead on general descriptions, such as "blue-white perfect," "slightly imperfect," and "commercial." Selling was much easier that way. There was no one judge who would determine the category that a diamond fit into. Today the business is "dried-up," he said.

Grading standards from the eighteenth and nineteenth centuries had used widely shared descriptive terms, which made the diamond industry a "tower of Babel." As one source said, "There was no way for people to communicate effectively. We had no common language" ("Labs," 1999). Other early terms included "first water," "Wesselton," and "cape."

Objectivity and standardization have been the goals of the GIA, but these standards remain elusive. The system grades the diamonds according to color, clarity, and cut. The color grades of D to Z are used for relatively colorless diamonds; another scale was devised for strongly colored diamonds. Clarity grades rate the internal purity of diamonds, and other grades rate the qualities and proportions of the different cuts. However, judgments about all values, except for carat weight, remain subjective, and individual skill remains necessary. The evaluation of rough diamonds is even more subjective.

Though the GIA grading scales are in almost universal use, the International Diamond Council's system is preferred in Europe (Shor 1993:229; Morrison personal communication, December 4, 1998).

Furthermore, the grading reports of one lab are not always consistent with those of another lab. In addition to two GIA laboratories in Japan, for example, I heard from dealers that dozens of other labs, each inconsistent from one another, exist in jewelry shops and elsewhere in Japan. Tougher grading standards that went into effect in 1997, however, were expected to standardize the Japanese market ("Japanese Market Expecting Big Changes in 1997," 1997:32).

Moreover, each country has different standards related to cut that reflect cultural preferences in aesthetics. Consequently, this aspect of diamond rating is unlikely to be standardized. GIA President William Boyajian said, "It's not an easy process because there are different (aesthetic) standards in different parts of the world, and the grade would have to take all into account" (Shor 1993:229). For example, some American jewelers prefer the cut devised by Marcel Tolkowsky (the "ideal cut"), while most Europeans prefer different proportions, and the continual introduction of new shapes exacerbates the problem of deciding on one standard. What, finally, makes one diamond superior to another?

"Imponderables always remain," a dealer said. Various components of the stone—the proportion, the "shoulder," the visual balance, the color, and the "life"—have to work together impeccably in order for the stone to command a high price. In addition, some fancy shapes can be more attractive than others. One dealer explained the evolving emphasis on cut. Since the acceptance of GIA standards, he said, the make has become much more important than it used to be; GIA instruments can determine exact proportions and other precise measurements. Still, cultural preferences remain. The Japanese want these proportions to be exactly right; others want different specifications, he said.

New shapes are always being tested and developed. Early in field research I went with a dealer as he showed a lasered stone he called a pineapple to his usual rounds of clients. Their exchange illustrates the mixed response that a new shape can evoke:

"Do you like pineapples?" the dealer asks his prospective customer as he tosses the parcel paper with the special diamond in it. "Yes," says the other. "How much for the pineapple?" he asks, peering at it with his loupe. "Twenty-two hundred per carat," says the dealer. "For a pineapple?" quips the other, turning it over and looking again. "How much would it be at Waldbaum's?" "Maybe a little less," replies the dealer. "$2200 per carat, for a pineapple!" muses the customer, still examining. The dealer argues, "This isn't apples and oranges, after all." The customer enjoys this, stands up abruptly, and decides to show it to others in his firm, wondering out loud, "Do we want to promote this?" as he folds it up and takes it to his brother's office next door.

After an industry study found that various kinds of cuts produced equally pleasing results, the study concluded: "It is our opinion that any cut grading system that attempts to categorize diamond appearance is premature in absence of a more complete understanding of the factors that give rise to this appearance" ("GIA Sets to Work on Cut Question," 1999). But as the "make" or cut become increasingly important to the New York manufacturing industry, the fact that the GIA does not rate cuts per se is a serious lack.

Creating an "ideal" and educating the consumer via grading standards and reports, however, also generates a paradox: "On the one hand, the trade wants, and even needs, to make more money for very fine-cut stones. On the other hand, if they educate consumers too much about cut, how will they ever sell the stones that are not cut so well?" ("Labs," 1999).

The GIA laboratory in New York resolves most gemstone identification problems. Spectroscopes that detect enhancements, computers that calculate the mechanics of cut, and other instruments that reveal artificial treatments are inaccessible to the ordinary trader. Though traders can often get by without using the GIA lab, its instruments can usually provide answers about chemical composition, imperfections, and artificial treatments not seen by loupe.

I heard two traders wonder whether a stone had been irradiated. The Japanese buyer and the American dealer couldn't determine whether the stone was natural. Since neither was sure, the buyer wanted the dealer to pay the GIA's fee to find out. They bantered and joked, spurred on by the fact that I was their audience. The buyer said he would pay the American back if the stone turned out to be natural. They continued bantering as the buyer sealed the envelope, but the American predicted to me that the Japanese buyer would not buy the stone.

Another afternoon at the DDC, a young trader approached an elderly dealer and asked him to examine a stone. The young man sported an earring, a somewhat unusual look in the DDC, even for the 1990s. He thought the stone was probably too good to be true, and he wondered whether the stone had been treated. The dealer took a cursory look and said, "Yes, it smells treated." "Four thousand dollars?" asked the young man. "No, more like $1800," responded the dealer. The young hopeful left, looking deflated. An observer said to the dealer, "That's the first person you've made unhappy today. Who's next?" The dealer responded, "He's young."

Experienced traders can make educated guesses about whether stones are treated, but the lab gives more definitive answers.

◆

The GIA states that its "Gem Trade Laboratory Diamond Grading Report has become the benchmark for the industry and can often be seen accompanying beautiful diamonds" (GIA website, December 4, 1998). Many consider the report to be the fifth "C," because of the clout it carries within the trade and among consumers. Though usually referred to as "certificates" in the trade, the reports do not carry guarantees and used to be issued only for large, fine stones that were difficult to classify and grade. Today, the majority of diamonds larger than one carat are graded. The one-page computerized report shows a schematic of the stone with mathematical ratios of the shape and a chemical analysis of the makeup, including whether fluorescence is present. It grades the stone according to the GIA system, complete with comments such as "Significant graining is present."

The increasing reliance on the grading reports has become a form of insurance for consumers, jewelers, and traders. A diamond that has a GIA grading report—and more of them do—sells for a higher price than a comparable stone without a grading report, as though the report certifies the stone's worth. As the GIA website states, "While the acquisition of a diamond is the ultimate symbol of love, it also represents a major investment in time, energy, and money. You need and deserve to feel confident in your decision and in the integrity of your stone. That's why having a GIA Gem Trade Laboratory Diamond Grading Report with your diamond means peace of mind." Further, says the site, "Whenever you buy a diamond, request that it be accompanied by the [report]. Ensure that your diamond has been graded by the pioneers . . . with the utmost impartiality." The message adds, "No valuation is stated." The impact of the certificates has grown as more people have used them.

Though the GIA standards mean that more details about a stone can be precisely calculated, judgment and experience remain important qualities for a trader. Furthermore, the reports should not be considered binding or definitively objective. An industry official told me he would always be careful to call them GIA grading reports, not "certificates." The GIA can't be successfully sued on the basis of the grading reports, he explained. Because humans, not computers, ultimately grade diamonds, the diamonds are evaluated according to subjective criteria—except for carat designation, which is measurable weight. He told the following story to illustrate: A celebrity gave his wife a D flawless, "perfect" diamond accompanied by a grading report attesting to it. When the wife had the diamond appraised some years later, the new grading report now gave the stone an E rating. The celebrity's wife attempted to sue but was unsuccessful, proving the point that the grading report does not guarantee the rating. In the final analysis a diamond is only worth what people will pay for it.

Many old-timers in the business think the ubiquity of grading reports threatens the art and judgment of a trader's expertise. Traders evaluate stones differently from one another, and some of them become known for a "good eye." Experience from seeing thousands of stones enhances one's reliability. Highly esteemed traders can be counted on to judge a stone's color, to spot defects in clarity, and to assess cut. When questions arose in the past—as they do today—traders conferred with one another about cut, color, the possible mine of origin, and other considerations. In the past, discussions of a stone by knowledgeable traders and manufacturers were an integral part of daily business. Today, one can get by without these abilities because the GIA is a reliable support—or crutch, as the case may be. Increasingly, traders know less about diamonds than about the world of buying and selling as more of them take GIA courses and fewer of them serve as apprentices to learn the manufacturing process.

A manufacturer takes a calculated risk when he or she decides how to cut or recut a stone and depends on his or her judgment as to whether the clarity, color, or cut would be improved by polishing. The downside is that cutting sacrifices some carat weight. A better GIA grading offsets the risk of recutting the stone and diminishing the weight should the new cut improve the clarity or the proportions of the stone.

As a dealer said, "Some of these things were manufactured twenty years ago. If I have a 2-carat stone, an SI1, [and] you can recut it to a carat-85, VS1, you're coming out very well. Things have changed in the last five, ten years. Certain things became more important than weight, some less important." A dealer told me about his friend who took great risks recutting stones. "There are all these tricks, and he knows all these tricks. He knows hundreds of cutters. This one specializes in this aspect, and this one can do that. Like a surgeon." However, in addition to the chance that a stone might not survive the recutting and the verdict from the GIA might not be what he wants, resubmitting a stone to the GIA takes time, and time is money.

"And then of course," continued the dealer, "after he's done all the work, the fact that it takes time to recut it and then put it into the GIA—the GIA takes a month and recutting can take a few weeks; it depends on how big an operation it is—then he has to go through the same worries of: Can he sell it? Will he get paid?"

One dealer, for example, resubmitted a stone that several dealers considered flawless that came back with a lower grade, indicating imperfection. He then considered resubmitting it again to try to improve the grade. Another time the partners discussed whether to do more work on a stone that the GIA had rated VVS1 and "improvable." They disagreed about how to proceed, then decided to "take some of the brown out" to improve the color grading. I went to the GIA with one of them. The dealer asked the GIA to confirm that the stone was indeed improvable. The senior member of the firm explained that the GIA had once downgraded his yellow stone a full color grade on resubmission to the lab. To contest this ruling, he brought in another yellow certified by the GIA to compare the two, and he "won out."

He was now prepared to fight if the GIA kept the rating at the H that it originally set. This time the decision to improve the stone paid off: the GIA upgraded the stone to a G. The firm estimated that the higher grade added as much as $10,000 to the worth of the stone.

The GIA admits a mistake by not charging if it gives a new rating to the same stone. If the rating does not change, however, the GIA charges half price for the rechecking.[1]

Some dealers believe that the GIA gives a tougher rating to a stone already rated than to one it has never rated, so resubmitting stones is done with caution and strategy. A dealer told me he hoped the GIA's computer would consider his resubmitted stone a new stone since its measurements were different after the new polishing and it might not be recognized as the old, already graded stone. The verdict that the GIA renders then significantly affects the next steps dealers take. For example, a dealer decided not to join in as partner on a particular stone with another dealer because the GIA had given the stone a worse rating than expected. Dealer A told me that Dealer B had "gotten tired" of the partnership and of his confidence in the stone; he wanted to back out. Calculations about profit and risk were now different. Dealer A bought out Dealer B for $500 to ensure that a good feeling remained between them.

Complaints about lack of consistency among labs persisted. I was told by several dealers that in the 1970s before the GIA's California lab got a computer, it was considered the more "lenient" lab in the United States to send stones to for grading. After the GIA's computer in California was up and running, some New York traders considered the two labs comparable. But one manufacturer who specialized in a particular kind of fancy colored stone told me he didn't consider the GIA consistent and said he could always get another reading. A trade publication carried this letter of complaint about perceived differences between Antwerp and New York labs: "[There is] an inability of the different laboratories to determine the actual clarity of the stone. It has come to the point that it is common practice for Antwerp dealers to come to America, purchase polish[ed stones] and recertify in Antwerp for a better grade, after which they sell it back to our own customers at a slight discount to what we must sell a GIA stone of a comparable quality. It has created an intolerable situation where, when I call on a customer to sell a JVS1 stone with a GIA certificate, I cannot compete if he happens to have a stone from Antwerp with an Antwerp certificate. Don't bother telling the customer that there is something more to our certificate. To your average jeweler, a JVS1 is a JVS1. Besides, to them our customers, as we carefully kick ourselves in the behind, it sounds like sour grapes anyway" (Twersky and Myerhoff 1993:2).

---

[1] In 2001 the GIA charged approximately $98 to evaluate a one-carat stone.

A dealer told me that the inconsistency in grading the same stone within the same lab can usually be attributed to different gemologists doing the ratings. Or, he added conspiratorially, maybe bigger firms receive more favorable ratings because they have more influence? Another dealer flatly pronounced this notion "groundless."

Lack of uniformity between labs is one difficult problem. Different aesthetic standards reflecting cultural diversity in taste is another; and lack of consistency in the grading of one stone submitted to the same lab more than once reveals perhaps a deeper flaw in the system.

As the walk-in GIA lab in New York is increasingly utilized for daily business, some dealers call ahead to see how many people are waiting to see a gemologist. On any given day the GIA is crowded with people waiting to receive reports. They strategize the right time to go. A broker kept calling to find out how long the wait was, how many people were ahead of him. Then when we went over, there were still nine people ahead of us. Too long to wait, we decided; later when we returned, the wait was reasonable. Delays are frustrating and jeopardize sales.

In addition, the turnaround time for receiving reports on the diamonds seemed to grow longer, and this trend caused annoyance. One dealer was disgusted that the GIA still hadn't returned his stone. The GIA had promised that the stone would be ready the previous afternoon. The dealer's $100,000 deal was contingent on a GIA report confirming its judgment on the stone two years ago. The dealer needed the report right away. Because the GIA's standards have not always been consistent, especially for fancy stones such as this one, a confirming report is necessary in such cases. The dealer told me that since the GIA has made more gradations in the categories for fancy colored stones, the grading has actually become somewhat more lax, so getting a higher grade is actually a bit easier now. Finally, the next day, the GIA's report validated the earlier one, and the sale was completed. The sale could have fallen through, however, because of the delay.

Another dealer called the lab to find out if the report on the stone that he had submitted the day before was ready. He was told that since he brought it in after 10, it wouldn't be ready until the afternoon. "We pay good money for this?" he commented, irritated. His partner commented icily, "So we'll watch this turn into sixty days because of the GIA." Being made to wait meant that their deal would be delayed and that they would be paid later; they were in effect extending credit to their buyer. One of them would go to the GIA to tell them this kind of wait was simply unacceptable; too much was at stake.

As the GIA has become entrenched in the industry, a new niche has been created. Some manufacturers now specialize in evaluating and improving stones for traders prior to their submission to the GIA. We visited one of these men in his tiny office to pick up a few stones he had evaluated. In a cubicle the size of a closet the man was stooped over his long, narrow, and grimy desk covered with parcel papers, memo pads, photos, tweezers, loupes, microscopes, and other paraphernalia, looking through a loupe at a stone. He turned around in his squeaky chair to speak to us. He stamped one of the parcel papers "OK to GIA" and told us what he thought needed to be done to the others to improve their rating at the GIA. By inexpensively removing some superficial flaws, he said, he enables a stone to make a better impression, which could result in a better grade at the GIA.

Another man runs a similar business from one of the long benches at the DDC, avoiding the overhead expenses of a private office. He charges a certain amount per carat to evaluate the stones. He told a man that his stone had some "nicks." It might come up to a VS or a VVS, but it might not be worth the trouble for the price it was going for, he added. Hearing that, the dealer couldn't decide what to do.

As fancy colored diamonds have become comparatively more expensive than white diamonds, the GIA has been doing more business in them. The GIA started grading fancy colored diamonds in the 1970s and changed the grading system for them twenty years later. I saw traders seek out those among them considered experts in this growing field for advice about the worth of the stones. Leeway still exists in the categories, and the grading system evolves as it reflects the trade. A dealer explained that as traders start to talk about different intensities of the colors "that seem to flow out of the stone" and trading reflects distinctions between colors, the discussion about the colors leads to the development of new names. The GIA has begun to assign names to the categories that the trade is identifying. These names then objectify and standardize those distinctions. Just as the old category "blue-white" was replaced by several GIA gradations, such as VS1 and VVS1, the GIA is objectifying certain colors with special names, such as "vivid." As it uses emic categories—that is, categories that the natives (the traders) consider valid—they are considered trustworthy in the trade.

However, complaints of inconsistency in the grading plagues the color terms as well, and the GIA does not dispute indeterminacy in the color grades. Though the GIA uses special lighting equipment for these stones, a dealer commented, "This is a subjective business anyway. There's no way they're going to make this a completely exact science" (Shor 1996c).

Many credit the GIA with opening up and enlarging the business, benefiting the trade overall. Its growing importance in the 1970s reflected the surge in diamond prices that decade, and the increasing interest in colored diamonds in the 1990s expanded the market. The process of categorizing and naming finer distinctions among cuts and colors and assigning them new value is likely to continue as more market niches are thus identified.

A dealer told me that the GIA was good for business. People who say that it's taken away business and that they can't make a living anymore are full of baloney, he declared. On the contrary; it's opened up the business, he said. With certificates going all over the world, a Japanese can study a certificate and see the color, carat weight, clarity, and cut; he can check out the make and have essentially no surprises when he decides to order it. So the certificates have helped expand the business throughout the world as people fax them to Singapore or Bangkok and elsewhere.

Standardization has brought increased "transparency" to the goods. The growing clout of the GIA grading standards and reports has helped demystify the business as it has brought almost uniform information to customer, jeweler, and wholesaler alike. But this trend reduces the profit margins along the buying and selling chain, and many middlemen in the system have been choked off and put out of business as a result. As more people along the pipeline have become knowledgeable about diamonds, they have become warier. A dealer commented that being better informed, customers are less trusting and more insistent on having the grading reports. "They're afraid they're going to be ripped off."

Acceptance of GIA standards paved the way for the Rapaport price sheet which could assign prices to those categories. The GIA grading standards and the grading reports have helped turn the diamonds into a commodity.

Stones with certificates automatically command higher prices and are traded with narrower margins than stones without such pedigree. A diamond without a certificate, however, can attract interest from more traders because it offers more room for negotiation. With the toughest bargainers, which the Japanese buyers are reputed to be, dealers said that they especially like uncertified stones because they can offer less on these.

While the GIA has helped standardize and commodify the evaluation and grading of gem diamonds, the human factor nonetheless remains. As long as large and fine diamonds are prized for their individual characteristics, much like art, antiques, and other treasures, the march toward uniform grading will always be incomplete. The ratings ultimately reflect preferences that in the final analysis are a clumsy reconciliation of differences in cultural and aesthetic standards. They remain rooted in an ultimately arbitrary codification of subjective judgment. The system of grades and ratings contains ambiguity around which argument swirls and from which haggling rules. Gaming the system—or using strategies to acquire the best ratings from the GIA—may have become more sophisticated as room to maneuver within the narrowing playing field has become more constricted.

◆

Once the GIA described and categorized goods according to a more or less uniform and widely recognized benchmark, a standardized price list attached to those categories was soon to follow. This added standard prices to the information flow in the industry. In 1978 Martin Rapaport began to publish his own price lists. These lists won worldwide recognition over competing price lists.

Dealers and brokers refer to the list and negotiate discounts in relation to particular cuts, weights, and clarity grades. On each two-page list Rapaport states, "Prices in this report reflect our opinion of HIGH CASH NEW YORK ASKING PRICES. These prices may be substantially higher than actual transaction prices. No guarantees are made and no liabilities are assumed as to the accuracy or validity of the information in this report." His website notes: "The Rapaport Diamond Price Sheet is considered the primary source of diamond price information by the jewelry trade. It is used in diamond markets world-wide as the basis for establishing inter-dealer prices—providing invaluable information for trading stones effectively and profitably. The price sheets serve as a starting point for negotiations and as a basis for establishing value for a broad range of diamonds."[2]

The price quotes reflect Rapaport's assessment of current asking prices based on international information and serve as a high-enough "initial price level which everyone can agree on" (Rapaport 1995:3).

Rapaport's impact on the trade has been huge, and the industry saw a threat from the start. The WFDB passed a resolution banning member price lists in 1982, but the First Amendment made it inapplicable in New York. Around the same time, the New York DDC initiated an antitrust suit against Rapaport, alleging price-fixing; the DDC also attempted to have a Jewish court prohibit his involvement in the Jewish community and tried to expel him from the club. As these actions coincided with the diamond crash of that time, many diamond dealers blamed his price sheet for their ruination. The price sheet reflected, not caused, the change, however. The price list also revealed the lowered diamond prices after the crash (Shor 1996d).

The uproar that the DDC action created backfired on the DDC. The Federal Trade Commission initiated a full review of the DDC for possible restraint of trade against Rapaport (Bernstein 1992:139n50). Rapaport sued the DDC, alleging restraint of trade, when he was kicked out of the club. He was readmitted after a closed settlement. One dealer told me that the DDC board had asked his opinion about what to do, and he had told them to settle—but his advice wasn't taken at the time.

After Rapaport won his case, he was elected to the DDC's board and later

---

[2] Rapaport also publishes a weekly magazine called *Rapaport Diamond Report* and has an Internet site called Diamonds.net with articles of current interest; the site is a secured site for buying and selling among members.

moved to Israel. Many dealers believe that this lawsuit made the DDC forever wary of kicking anyone out of the club—a stance that may have further decreased the DDC's clout.

Many blame Rapaport for fiercer competition and say that diamond margins have been reduced by half as a result ("Rapaport Fancy Prices Provoke Protests," 1996). Two dealers with fifty or more years' experience in the trade scoffed with disgust and said that the biggest changes in the industry were the (GIA) certificates and Rapaport's list.

"Now we have to memorize their lies instead of our own judgment," one of the two dealers said. "It's a big psychological change and it hurt the business. There was more mystery before. People used their judgment. The list confuses the public, and it's forced some of the small men out. Rapaport basically gets his prices by asking several men their opinions and that's the price he sets. Maybe the DDC should have used propaganda against him rather than just throw him out," one said. "The Rapaport prices are a fiction. If you go by these prices, you're licked. You still use your own judgment when you discount from the list."

Another said, "In the beginning he had a price list, and in '80, when the price began to fall, he would amend the list to reflect the market conditions—and the price of course went much lower, and they wanted to lynch him as a result. Now he's become smarter. When the price fluctuates, he leaves the price alone and changes it very gradually over a long period of time—very minute changes—and if there are any differences in price people quote the offer, let's say 60 percent of the list, 65 percent of the list, 58 percent of the list. The list is still the norm, and the quotation becomes a percentage of that list. But still his price is still the thing people go by."

Traders generally copy the weekly quotations and share them among themselves, using the prices as a starting point for discounting certain percentages depending on the gem. As one dealer noted, "The Rapaport list is an accepted convention which is not taken seriously at face value because . . . it's a conspiracy against the public essentially. So there are discounts from that list. In the old days the jewelers knew very little of the price, so some dealers made very high prices with some of the retailers. But now that they have the list, everyone knows the price. This sort of normalizes it."

Rapaport has written, "The benefit of the Rapaport Prices is that they give the industry a common denominator or standard base level from which to begin negotiations and allow for a standardized method to communicate price information" (Rapaport 1995). Some dealers maintain that the price lists have expanded business but narrowed the profits.

Certainly, the public, not just the retailers, is getting more savvy. In this business, as in most others, consumers have become more sophisticated; increasingly they go directly to the source and have more information about products and prices. One dealer noted, "There are some shrewd ladies looking for bargains, and they come to retailers on 47th Street armed with the Rapaport list, and they say, 'What do you mean, charging me 30 percent below Rapaport! I can buy for 40 below Rapaport!'"

Despite the criticism that Rapaport has caused narrower profits and greater commoditization, Rapaport denies an adverse impact, contending that grading standards and his price sheets provide reputable information to an increasingly discerning public. He says the accuracy of the information helps honest diamond traders, exposes dishonest traders, increases the trust between buyers and sellers, and thereby benefits the trade overall ("The Truth about Profits," 1998). While a KVS1 used to have a price judged by a trader, the KVS1 is now assigned a standard price from which negotiations begin.

When a new list comes out, traders in the DDC and New York offices see how the prices listed reflect their knowledge of the trade. One day in the club I watched several dealers and brokers discuss the new list. One broker asked another if there were many changes from the last list and was told yes. Significant changes in prices of the fancy shapes seemed to surprise some of them. Quite a few prices were higher. They disagree with the list a little, saying that some items seemed overvalued. Some of them concluded that the market must have become firmer. De Beers is tightening the supply, which increases both demand and price, the men said. The 5-carat didn't go up at all on the list, and two other men discussed this.

Though the list is supposed to reflect the prices that are actually traded, the relationship between what Rapaport reports and the results of the report is more complex. For example, once in the mid-1990s Rapaport raised prices on various fancies but provoked bitter complaints from traders, who said the increase was not nearly high enough ("Rapaport Fancy Prices Provoke Protests," 1996). Some argued that Rapaport was destroying the business.

A dealer complained about the incident that had occurred a few weeks earlier: The list had come out, and the 5-carat J pear shape VS was down for 30–35 percent below list. The emerald and radiant cuts were more abundant, and they were being traded at 40–50 percent below. Rapaport was "against this," the dealer contended. So four weeks ago Rapaport had come out with a list for each of four shapes, pears, marquises, emeralds, and either heart shapes or ovals—he wasn't sure which. In this new listing Rapaport reduced the emerald cuts 20 percent more. He changed the lists so they were much lower in order to catch up to the reality of the marketplace, but he ended up skewing it and creating cockeyed prices that didn't make sense to anyone, the dealer said. "Everyone rebelled" and threw the lists out, he added. Rapaport finally admitted his mistake, the dealer said. To reflect what the market is doing is one thing; that's what he should be doing, he added. But to dictate what the market should do is another thing entirely, and that's what he tried unsuccessfully to do, he said.

Another dealer summed up, "He's missed the boat this time and lost his credibility. He's one reason why profits are so low." Another added, "Now he's making sure we don't have any." Though Rapaport's disclaimer in that week's price report—that well-made stones should carry a premium and less well-made stones should be considerably discounted—accompanied the lower prices he listed, dealers were apparently not satisfied, *New York Diamonds* noted.[3]

A flamboyant and controversial character in an industry historic for its share of characters, Rapaport has the reputation of being a genius. Some blame him for their difficulties, but many believe he is an apt reflection of the times. In a conversation I overheard about the predominance of Jews in the industry, someone asked if they could think of any WASPs. Someone immediately named Martin Rapaport because "he stings."

A truism about the marketplace says: "Those with the least knowledge pay the most" (Shor 1996d). The more information a trader has, the better off he or she is. Rapaport says diamond pricing is democratized because the GIA standards and the Rapaport lists provide more information to more people. But people argue about the impact. A dealer told me, "All around it may be better. It's better to have an equal standard. It's better for business as a whole, I think. It smoothes the business as a whole—gives more stability maybe, less fluctuation. And people have a sense that there's a certain price written down somewhere. Probably gives them a sense of security when they buy it. It's my own opinion. I might be wrong about this, but I think it might be better for business as a whole."

Another dealer said that, in a way, Rapaport made the business easier because people could know the prices—but unfortunately, making the prices so public also makes it hard. If the people in the garment business told everyone how much it took them to make an Armani suit, he said, how would a place like Bergdorf make a living? There's no room between wholesale and retail. He contended that Rapaport didn't care about the impact, because he made so much money by selling the price list to so many people.

◆

In early 1999, the large diamond firm of Lazare Kaplan concluded a ten-year exclusive worldwide marketing agreement with General Electric to

---

[3] *New York Diamonds* reported that when he increased the prices on his list the following week, he explained, "I didn't accurately reflect the premium prices paid for finely made fancies in the New York market" ("Rapaport Fancy Prices Provoke Protests" 1996:36).

begin marketing diamonds treated by a new and undetectable permanent "enhancement" process. Approximately 1 percent of diamonds were potentially treatable. Lazare Kaplan's new subsidiary, Pegasus Overseas, would perform the enhancement. Shock waves penetrated the diamond industry.

The concerned dealer who told me about this said everyone was talking about it, and what would happen next was unclear. This was unprecedented, and the implications were enormous, he felt. He was worried that confidence in the industry would diminish significantly.

According to Rapaport's website, Lazare Kaplan was buying large quantities of brownish diamonds to be enhanced by the new process and to emerge lighter, whiter, and more valuable ("Rapaport: Trade Alert" 1999).

Two issues in particular alarmed the diamond community. One, because the treatment process appeared to be undetectable by GIA lab methods, the diamond industry wanted complete disclosure of the process so consumers would be informed of artificial treatments performed on stones they might buy. While Lazare Kaplan contended that a permanently undetectable treatment needed no disclosure, critics insisted future technologies might reveal the treatments.

The second concern was the reputation of the industry in general. Wouldn't consumers lose confidence in all diamonds if they knew that the diamonds they were buying might have been artificially treated? Rapaport wrote at the time, "Interestingly, if LKI [Lazare] is correct and the treatment process remains forever undetectable, then a class of diamonds will enter a grey area of uncertainty. LKI will argue that the diamonds should be priced and sold by their apparent color and that the treatment is irrelevant since it can never be detected. GIA will argue that if they do not know if a stone is color treated or not they must say so on the report. The issue is this. If a treatment is undetectable is it a treatment? GIA will say yes because at some time in the future it may be detectable and therefore needs to be differentiated from the general population of stones. LKI will argue that since no one can tell that the stone has been treated, for all intents and purposes it is not treated" ("Rapaport: Trade Alert" 1999).

New treatments to enhance the color or clarity of diamonds are normally developed in collaboration with the GIA—but not this time. The GIA president commented, "GIA has not been privy to the development of this alleged treatment, and to date has received very little information from the parties involved" ("GIA Calls for Facts on Purported Diamond Treatment," 1999). Meanwhile, GemStar, a company in Illinois, claimed to have developed a similarly undetectable process that cleans stones and improves color permanently (though efficacy of the claims was not clear). The secret process was called GemTech, and the enhancement was to be performed at the mine.

Alarmed questions swirled. How would business fare? Wouldn't confidence in the industry be undermined? Some contended that the process would not legally need to be disclosed if it were truly irreversible. A jewelers committee argued that the lack of detectability of the process made disclosure more important, not less. An opinion survey on the Internet collected 323 responses, only nineteen of which supported Lazare's right to nondisclosure. An Australian noted, however, "'Sometimes I think you Americans think that diamonds are some kind of cure for cancer or a part of the space program. After all is said and done, it is simply another commodity.'" (Nestlebaum 1999:1).

If disclosure were required, some said, dealers who disclosed the treatment might be harmed while dealers who did not disclose the information might benefit.

By late 1999, the GIA, GE, and Lazare had reached an agreement. They decided that all GE-processed diamonds would be inscribed with "GE POL" to indicate that the diamonds had undergone the special process. The GIA was researching the process diligently. Expressions of approval and relief were widespread. Nonetheless, the Pegasus incident left the diamond community shaky and uncertain about future developments.

◆

The last few decades have seen the introduction of the personal computer, overnight delivery services, the facsimile machine, electronic mail, cellular phones, and the Internet, all of which have increased efficiency and have enabled trades to go from manufacturer to customer less encumbered by middlemen. This momentum has helped large firms get larger and has helped others find a niche.

Electronic buying and selling via the Internet is affecting the trade, though not as profoundly as the GIA grading reports and Rapaport's price list. On the web consumers can locate jewelers, and they can learn about gems, grading, prices, and the like. Companies use the web to sell directly to consumers and to one another. A dealer told me, "I'm a member of Polygon and it's a bunch of wholesale dealers throughout the country—actually sometimes overseas also—and they have a particular call, they request a certain kind of stone, and they post it. And they can get a big response if people have it or no response at all. One of the biggest problems is that they're usually looking for things and then they write their budget, and they can't afford what they're looking for." His partner noted, "That's the biggest problem in all the calls that you get, [through the Net] or through the telephone." The dealer protested that good developments sometimes flow from the Internet anyway. In this particular example, he continued, "I recently made two contacts in New York [through the Net], and one of them I made a sale."

After he gets references and checks them out, business can proceed. His partner commented, "We make contacts normally without the computer. Somebody comes to us with a request that we don't know, so we do the same thing with them. We check them out. So this is just a new avenue for us."

Some jewelry companies were quick to put their web pages on the Internet, and they were immediately joined by Jewelers of America, the American Gem Society, trade magazines, the GIA, and others.

Rapaport says the information systems provide opportunities that will lead to greater specialization ("The Truth about Profits," 1998). Though some think the clubs are the best places to trade, others note the shrinking role for the middleman as retailers and manufacturers deal directly using faxes, the Internet, and couriers. Web selling has enhanced consumer knowledge, accelerating the push for lower prices and increasing the squeeze on middlemen, especially brokers ("Diamonds on the Internet," 1995). Though retailers have begun to use the Internet, they have derived little benefit from it thus far (Nestlebaum 2001). The possibility of scams and credit card problems has been one downside. In 2000 Rapaport noted that diamond jewelry e-tailers were still very unprofitable as they spent too much on advertising ("Diamonds.com," 2000). Though Internet retailers sell at a discount, the threat for the wholesaler has been less than what was feared, according to the *New York Times* (Weber 2001:11).

Overall, companies get bigger and more efficient. Whereas rough gem diamonds used to travel from sights to smaller manufacturers to the dealers and finally to the retailer, now more wholesalers deal directly with the retail trade. Whereas most of the business used to be for consumption in the United States, the bulk of business is for the overseas market today. Zales and Kmart have become big sellers of diamonds. Meanwhile, Harry Winston, a prominent diamond firm known for luxury jewelry, produces jewelry for mass marketing in China, and countries in East Asia and India polish vast amounts of tiny diamonds for jewelry for mass consumption.

Cell phones are a constant presence now in offices and the DDC. In addition to the intercom calling individuals to club phones, individual cell phones ring incessantly. Traders walk the streets, talking into the phones against their heads. The deals are accelerated, creating intensified pressure.

Overnight services allow the stones to travel quickly. Fax and scanner transmit the exact specifications of a gem's grading report so the customer can determine whether he wants to buy. The fax helps jewelers provide instant insurance for customers on the jewelry they buy. Computerized trading showcases the merchandise with certificate and price information. Low-wage competition from Asia has spurred many firms to automate their manufacture. They argue that if automated manufacturing is done properly, it will provide a consistent product with increased efficiency (Shor 1993:

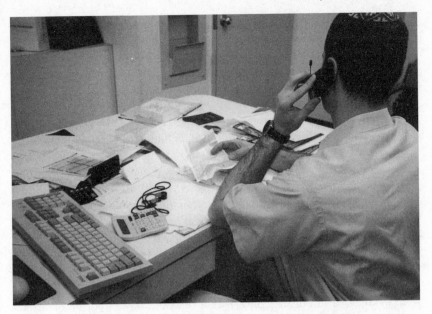

Modern Orthodox trader mixes the old with the new. (Photo by author)

226). Automated manufacture requires expensive equipment beyond the scope of most of the smaller and many of the larger firms, however. Some worry about the waste in automated manufacturing, as well.

A manufacturer of industrial diamonds liked the fax because instant pictures helped convey to customers in the Pacific Rim, Southeast Asia, and South America exactly how a diamond could be made to fit a particular industrial tool. With pictures going back and forth, the manufacturer could design an industrial diamond to exact specification.

However, despite the movement toward bigger operations and electronic assists, the New York diamond business still depends on face-to-face interaction. The stones that are traded are larger and more specialized, and the personal element in the business is still considered necessary. Though more people are going on-line, the traditional wariness and caution endemic to the business still ensure that traders deal with small numbers of people with whom they have trusting relationships. To open oneself up to trading with unknowns presents too great a risk for many.

◆

Meanwhile, the middleman remains squeezed. Rapaport says De Beers, not he, should be blamed for the tight market and narrow profit margins.

De Beers deals with large organizations; some dealers told me of companies that can buy $2–$3 million of diamonds a month.

As De Beers maneuvers in the complex global marketplace, it works to maximize its own position, less protective of individuals than it used to be. This reality was ever more stark as 47th Street shuddered from the news in 2001 that De Beers was going private and was going to compete with local businesses. One diamond dealer remarked, "They did a triple bypass. They're going from the mine to the consumer and skipping the middleman" (Weber 2001:11).

A manufacturer told me, "The middleman has been forced out for quite a while. There are less people, less brokers. There used to be quite a few women brokers. They usually took over when their husbands died, or if they didn't have something to do, and they had contacts with knowing a lot of people and they could work part-time and make a fortune, if you're good. But there are very few—in the whole city, if there's ten, it's a lot. There used to be hundreds. People [now] tend to deal directly one with the other. . . . The whole idea of brokering is out."

Individuals in the district have varied reactions. One told me her small business was doing well—perhaps because she adapted by going to jewelry shows and sponsoring foreign buyers in the club. A manufacturer told me he had to cut out the broker: "It used to be a manufacturer would give it to a broker, who would show it to a dealer, then the dealer went to Kmart. Now you deal directly with Kmart. Business has been 'sanitized,' and a lot of people who were undercapitalized fell by the wayside because they were bankrupt. One or two a year. It's tightened up. There are fewer people employed in the business. More can be done on computer. Inventories and distribution can be done on the computer."

Everyone in the business agrees that as a result of all these developments, each person finds his or her own way to make a living—or doesn't make it. Business has always been hard, but more small traders in the United States used to be able to get in on the act and survive. Survival of the fittest seemed a little more lenient back then.

◆ ◆ ◆

[7]

Diamonds, Families, and Time

Most of the wholesale diamond firms on 47th Street are family firms; most of the dealers, brokers, and manufacturers are male; and many of those who are able work into old age, sometimes well into their nineties. How do people start out in the business? Do families still encourage their children to enter the business? How do men and women view the women in the trade? How do individual traders regard their work as they age, and what makes continuing to work into old age attractive? Does working into advanced old age in this urban "life-term social arena" (Moore 1978) dramatize the advantages of staying active and social with the same group of people?

Many wholesale diamond firms in New York were begun by a father or a pair of brothers who recruited other family members, usually male, to work with them in some form of partnership arrangement. Most of these firms were started one, two, or three generations ago, often in Europe. In some cases wives worked beside husbands, and daughters and sisters were recruited to help when the father or uncle died, became ill, or needed assistance. Usually women were secretaries. Over the decades the women either migrated out of the business entirely or went into the more female-friendly retail side of the business, where they have been increasingly welcomed. A sturdy minority of women entered the manufacturing and trading realms of the wholesale trade.

Most young Jewish men followed in the footsteps of their fathers when there was an established business. Though it is no longer true today, most diamond traders started at the wheel. Back then, one needed some money, a little business acumen, and a willingness to risk in order to start in this difficult trade.

How many generations do diamond families really last? *The Family Business Review* asserts that 70 percent of all family businesses are sold or liquidated before reaching the second generation (Lawlor 1998). Though some of the traders I spoke with claimed multiple generations in the industry, Herman, sixtyish, questioned the fabled generational depth of most diamond families and thought two or three generations were the extent of involvement for the majority. He ticked off a few prominent names, such as Willie Goldberg and Louis Glick, who were first-generation cutters, some of whose sons were continuing. On the other hand, Fabrikant, a large business, seemed to have no heir apparent. In other firms, he claimed, the second generation went into real estate or became professionals, the end.

Herman's own father came into the diamond business, and he has continued it. But his three children are in different fields. Herman thinks we romanticize continuity in this context. People came from *shtetl* Poland to Amsterdam or Antwerp, sure, he said—and children came into the business because the education that was generally available to Jews was religious or vocational, and then the next generation tended to drift out. In this country multigenerational family businesses exist, but they're a minority. In Europe, he said, maybe the father looks at his 18-year-old son and says: So, go do something productive—be a doctor or do this. In this country, he believes, parents are more understanding when their children want to do other things, even sometimes art history, he adds. He mentions more diamond families—thinking out loud to himself, testing his theory—and decides it holds.

Another dealer, Marcus, said he believes that the family-business aspect of the New York diamond industry is strengthening. His father came to New York in 1915 and was active in the club at its inception. A lawyer first before joining his father, Marcus has four children. Three of them went into diamonds. Another son has a retail store. Though some family businesses break up and experience conflicts that sometimes result in family members suing one another, Marcus wondered whether the quality of family relationships affects their continuity.

Small diamond firms are typical in New York. Conflict and split-ups are fairly common in the diamond business, as they are in family businesses in general. Sometimes the splits are amicable; sometimes simmering distrust can cloud—or sever—the relationship. Josef's story:

When he came to America from Belgium during World War II, Josef approached his cousins, but they rebuffed him because of a misunderstanding. During the war the cousins had sent him $30,000 to ship diamonds to them in return. But when he couldn't ship them, he sent them their money.

Now he learned that they hadn't received the money, and they thought he had taken it. In 1941, the Canadian government sent the cousins a letter explaining that the government had the money. So when the cousins learned this news, they approached Josef to apologize—but he had decided to have nothing to do with them. He related the story with pride.

Though many clients of Jewish Family Services prefer working in family firms because "in business dealings it's usually safer to trust a relative than an outsider," Leichter and Mitchell noted that mistrust of outsiders went along with trust between family members (1978:78). Despite the stated preference for family businesses, their study of many small single-generational family firm members reported conflict about hiring, termination, family obligation to the business, feelings of exploitation, and the blurring of kin and business relationships. Silverman also noted tensions in a family business even as it kept "kin close by, provided jobs during hard times, had a cheap and loyal labor force, and allowed Orthodox Jews to keep their traditional religious customs" (1988:170–71).

In general, however, people told me of positive experiences in their family firms. That they did not speak of conflict did not mean it was nonexistent or even rare, of course, and it is hard to imagine a family firm not having some dissension or resentment. Yet, people in family firms perhaps also learn to work together more or less harmoniously. Also, though dissolution can result, conflict is often enjoyed in Jewish families as part of a culture that celebrates argument, disagreement, and spirited debate. Furthermore, Leichter and Mitchell's sample may have had self-selection bias since the clients sought assistance from the social service agency.

◆

Many parents in the diamond business think their children should learn a trade or a profession other than the diamond business. In the United States they counsel their children to go to college, become educated, enter some kind of marketable profession. The family diamond firm is a fallback position should the other pathway become blocked or prove unsatisfactory. Elaine, for example, tells her children to train for other professions—perhaps one day, if they still want to enter the business, they can. Her husband had training in an academic field prior to going into the business, too.

Young and old, male and female—traders I spoke with started out in the business in greatly varied ways. These are some of their experiences.

Sol's maternal grandfather emigrated from the Netherlands in 1894, when a Dutch diamond firm sent him to establish a diamond-cutting factory in New York. Sol said he could trace the diamond-cutting heritage of his family about seven generations—he'd looked up birth certificates that identified the men as diamond cutters. His father used to tell the famous joke that a Jew in Holland could be anything he wanted to be—a cigar maker or a diamond cutter. Dutch Jews, one generation after another, became diamond cutters. It was very specialized work. It was unusual to know all the steps, so each knew only a few cutting techniques. Sol's grandfather knew them all, however, and was considered a "fantastic mechanic."

Josef, an eighty-nine-year-old manufacturer, began work as a jeweler in Poland "because my father was a jeweler." He and his family immigrated to Belgium when he was young. After his bar mitzvah at age thirteen, he wanted to go to school, but his father said: Not on my time. So Josef continued his education at night. In the old days, Josef claimed, parents were different. They were strict but didn't care about school the same way. Schooling should lead to a practical outcome. And "I did it the same way with my son."

Stefan, a diamond manufacturer from Europe, now in his seventies, wanted to become a doctor, but the war intervened. Stefan's maternal grandfather was a diamond cutter in Holland. Stefan's father was a diamond cutter who moved the business to Antwerp after World War I, when taxes in Holland became prohibitive. After coming to the United States in 1942, Stefan enrolled at MIT. When his father became ill, he asked Stefan to help out in the business. His father had trained him in Antwerp because it was "knowledge that no one could take away." Eventually, Stefan took over the business.

After college and the Army, and while attending law school at night, Henry worked in his father's business. It "won out" because it was varied, interesting, surprising, and even exciting. His mother recalled how her husband pressed him: "[He] expected it very much. He was not a very modern father. He didn't listen. . . . *My* father expected him to go to the State Department at least! And my father, in contrast, said, 'And now you're going into 47th Street?'" He was disappointed. "But then," she added, "once [Henry] caught up, yeah! He used to go by himself to find customers. He went into the Village, you know—into the little tiny stores. It was really going from the bottom—very slowly—and it took him quite a while to get in, and to get the spirit of it. But then, either he caught on or it caught up to him. He did like it. He did like the international flavor of 47th Street, the way you could use your languages or whatever skills you have—you could *use* it."

Some people surprised themselves that they ended up in the diamond business. Stu, an energetic, wiry man in his mid-forties, taught in a university, then his father asked him to work with him because he was having trouble with his memory. Not making much money teaching, Stu agreed. He'd always had a good relationship with his father, but when he came into the business, he found how well respected his father was. Though his father died only two years after Stu entered the business, Stu feels that he keeps his father's memory alive through his activities.

A young man with knitted *yarmelke* perched on his head, Asa, had a sweet face with an impish look. He said he considered himself different from the others; in fact, after having worked with his father for a few years, he had left to teach Jewish music. Five years after that, he was back—but not forever, he told me. One of the things he liked about the business was having his own timetable and privacy in coming and going as he pleased. Business is tough, he said, but he wished people would not take it so seriously. Collecting what people owed was becoming increasingly difficult, he said—and the diamond business is not a young scene!

Women, like men, had varied experiences starting out. Beth, a jewelry designer, married and studied fashion design. Then her father asked her, the only child, to join his business. He died three years later. She never regretted her decision and stayed on.

When Francine was very young, her father used to give her minor jobs like weighing the stones when she came to the office. His family came from Amsterdam and Antwerp. Not intending to enter the diamond business, she worked for a bookseller after college but without direction. Her father suggested she try the business for six months. (Two men in her father's office left because she was on board, however.) Surprisingly, it caught on with her, and she was good at it. First she sorted diamonds for one or two years. Then, when she knew what she was doing, she was allowed to deal with customers, and then to attend sights—first with her father, later by herself. If she has children, she said, she will encourage them to enter the business.

Phyllis had been an office manager in Israel, her children grown, when she decided to try the diamond business. She needed the money. First, she supplemented with office work in a diamond firm in Israel. Because she showed an interest in the diamond work, her employers taught her how to sort. From there she became a broker, then a member of the club.

Elsa, a woman in her fifties, had earned a doctorate, gotten married, had children and taught, before going to law school. After working for one high-powered firm after another, she became dissatisfied with corporate life. Her father, a diamond dealer in his eighties, then asked her to help him and her brother in the business; her father had been working fewer hours since having a heart attack. He trained her but then bossed her around, though technically he worked for her. He liked keeping control, she told me mildly. Still, she enjoyed the switch her life had made.

Joyce called herself a fourth-generation jeweler but originally had no intention of following the line. She was in a master's program in physics and was going for her doctorate. Through a series of flukes, she went to Europe, lost time in school, and met a diamond dealer who wanted to change the industry's perceptions of women. He told her, "I'll make you the youngest woman in the club ever." Her father was flummoxed. "What are you doing? I thought you were in academics!" he said. Though she had never liked the business as she was growing up, this was different. She could be respected as a woman. Her boss made her a member of the club, and her father guaranteed for her.[1] Her boss taught her the rough business, she took GIA classes, and she began to enjoy the business as she realized her sales ability.

I also collected information about three men who were deciding whether to go into the business. Ultimately, Simon joined his father's firm; Daniel and Lewis chose other work. Lewis's grandmother told me of his enjoyment the summer he worked in the office when he was seventeen. "When Lewis was starting to work for us, he did not have much interest in the stone, or in how it was made, or why it's a certain way. Lewis loved the drama of selling. If he didn't know somebody, he could make a contact. He was aggressive that way. He always made like a play out of it. It was something his grandfather never did, even if he would go out selling, but he never saw that part of the business. For Lewis, I think he must have thought at night what kind of a scene he would create. He would put on special clothes to provoke discussion. He loved that. He always wanted to become an actor."

His cousin, Simon, was seventeen in 1986 when I heard his grandmother suggest to his mother that he work in the business during the summer as Lewis had. Simon's mother thought he had a good eye and would do well. I asked Simon's father if Simon might join the business. I don't know, he answered, studiously noncommittal. Maybe another grandson, maybe Lewis, would do it. And granddaughters? I asked him, his daughters? No, he didn't think full-time work in the business would be a good idea for them. . . . No.

Simon enjoyed that summer. A few years later, after polishing and cleaving with his grandfather during summers and between college semesters, he entered the business. During one summer, he had cleaved a stone nicely, and as a reward, his grandfather gave him a $50 tip—fantastic! Simon's father recalled admiring his father when he was very young as he calculated changing money from Dutch guilders into Belgian francs. He wished he had his father's ease and confidence. For his part, Simon said he liked the math and the geology of the work. He had gotten married and had started grad school and teaching, but he didn't see himself as a professor.

His grandmother commented, "Simon wanted to learn how to manufacture, and we really don't have the factory anymore, and there's nobody that can teach him. That probably would have been a pleasure for [his grandfa-

---

[1] As stated in the DDC bylaws, a diamond firm may request one or more additional designees from the firm to become members of the DDC as long as they "submit a full financial guarantee" (Article III, section 2B) for those individuals.

ther]. [He only learned cleaving,] not the small handicraft of it . . . but the analysis, the logic of it, and the figuring out . . . [it] probably would have clicked. The angles and also the figuring out: What do we lose if we take this spot out? We lose some diamond; we lose some value. What do we gain? If we don't do it, maybe we lose even more because it's going to be a dirty diamond. He would have enjoyed it, and I think Simon would have enjoyed it, too."

Daniel was another story. In 1986, he was twenty-five and had just started in the family firm, after a job in health care. The only one of three brothers interested in diamonds, he had been invited to join the business by his father. He'd waited until his father asked him because the timing depended on how their business was going and how much time they could spare to teach him. He hoped to advance quickly. A month later he was proud of his accomplishments, having just "finished" his first half-carat stone. He'd started with ten-pointers (10 percent of a carat), but they were too hard because they were so small. It was a matter of getting the eye tuned. The first stone had been simple. The stone he was working on now was harder because the proportions were off a little. You have to learn to trust your eye, he told me. As he didn't want to mar the stone, he was asking his father and uncles for help, and he was beginning to feel less anxiety.

A month later I asked how it was going. He was on his sixth stone, a half-carat stone. Some weeks later, Daniel's father, Jim, said he thought Daniel was doing pretty well. But explaining how to see something in a stone was difficult when your student didn't see it in the stone, he said. It was a good thing that Daniel's uncle was there, he said. It's so hard when it's automatic for you, he said, but Daniel was getting there. Daniel himself said he was doing "all right" working on a pear-shaped diamond that was easier than what he had been doing. Finding the grain in the diamond was difficult after doing several facets. Losing the grain was easy, especially when it was off-center. Several weeks later Daniel reported that he liked his work on the pear-shaped diamond because it was less precise than the faceting that had to be done around the stone. Since his father was making the important decisions, he wasn't very worried about consequences. Jim told me a month later that Daniel was doing "OK."

Daniel stuck it out for three years and then left. According to his father, "He had no fire for it." Perhaps Daniel could have set up a European office for them, but it never happened. Maybe the eagerness for the business skips generations, he said. He and his brothers had the fire, but they don't see it in the current generation. Though they respected Daniel's decision, they were disappointed.

Other young men provided different perspectives. A man in his thirties, Louis, was not sorry that he had joined his mother in the business about seven years earlier. He had gone to college first, considered the stock market. His brother and his uncle had been in another business, but they had recently joined Louis and his mother. Louis had taken a GIA course first to get acclimated, and then he learned some cutting.

They had recently decided to do more manufacturing, he continued, bucking a trend in the other direction. They were partnering with an older man who wanted to work more part-time, so the arrangement suited all parties. In the mid-1990s Louis saw the industry getting tighter and more difficult, however. De Beers was letting fewer diamonds out at the sights, he said. He saw people going bankrupt. He saw the Israelis pushing credit crazily and taking amazing risks. He saw the Asian market falling, which slowed everything. Nonetheless, while people complained, there was money to be made, he said. Business is cyclical. People don't get rich, but they can make a living, he said.

Yankel, a broker around 27, was a Hasid with corkscrew *payas* that dangled from his angular face. He had come into the business two years before and had two good years, but business was harder now, in 1996. Though he had recommended the business to his friends, he wasn't so sure now. He didn't regret it, but he saw that there was little profit that he could make.

A young man from South Africa had no family connections but a passion for gems. Slight and intense, he started by sorting in a shop. He hadn't been in the business long, but he knew that he would not want to do anything else. In contrast to everyone who said that working in the business without family connections was impossible, he claimed no disadvantage in not having family contacts.

Others said the opposite. For example, Simon and I went to see a twenty-eight-year-old man, Sammy, the youngest of three brothers who worked in their father's firm. Everything—*everything!*—was contacts, he said definitively. Sammy occupied his own office and as we talked, various members of the family walked in and out. With a Romanian mother and Iranian father, the family spoke Farsi, Hebrew, French, and Romanian, in addition to the English he flawlessly exhibited with us. Learning the business six years earlier, around 1990, had been frustrating, he said. He wished he knew then what he knew now. He could already see changes and knew there was less profitability now.

Even so, he was glad he had entered the business, he said. There is a kind of "greenhouse effect," as he termed it, a warm and comfortable feeling inside the "greenhouse" that is his family. If you're careful, you can make it, he said. He didn't know whether he would advise others to go into the business. On the other hand, he said, things could get a lot better in a few years.

Likewise, three Hasidic men in their twenties who hadn't been in the business long stated that if you were young and didn't have a father in the business, you couldn't make it, period. The GIA course was a way to start, but dealing would be impossible without family.

At the office of a thirty-four-year-old man of Italian ancestry, elegant with rich tones of browns and sepias, Giorgio tossed a series of parcel papers of colored diamonds to Simon that Simon examined, appreciated, refolded, and hurled back—not biting this time. Giorgio joined his father's business after his father asked him to consider trying it as a way to supplement the knowledge and skills he had gained in real estate. Speaking thoughtfully, he repeated my question to himself:

"Why am I in this business?" he mused. "Because if I wasn't, the business would die with my father, simple as that." He paused to answer the phone and within a few seconds said, *"Mazal,"* and hung up. Deal done. You can be much more persuasive face-to-face than on the phone, he said. You lose a lot with the phone. This business is unlike any other, he added. Whereas you basically hear English in New York, you hear a variety of languages on 47th Street. He loves that. A new person coming into the business must be under someone's wing. You feel invincible and protected. That makes you feel as though what you're doing is simple, even "child's play"—in the beginning. But then you start to become aware of how intently people are listening to you, and of how many important mistakes you can make at any turn. Simon nodded vigorously, remembering. Simon added that a slip of the tongue could be disastrous to the deal. You make money when you buy, Giorgio said, not when you sell! Simon said his grandfather told him the same thing—in other words, buying goods shrewdly will result in being able to sell them well later. It's unlike other businesses in this way. That's the challenge and pleasure of it, they added. A young man can grow faster in intellect in this business than in other businesses, Giorgio continued, and he loves that.

Lipschitz described the way the Amsterdam trade was imbued in a child's life. "When my Dad and brother came home in the evening, we all had diamonds for supper! When I was a lad I was once at my father's office when a couple of Americans came in. They had brought a ten-carat stone. When they asked me to hold it I automatically put it on the back of my hand, because that's the way you're supposed to hold diamonds. The Americans were really surprised, but I'd seen that sort of thing all the time as a child" (1990:129).

Some of the American children of diamond families whom I encountered felt the same connection to the business. Francesca had left home to begin architectural studies. But she did not stay in school for long, because she realized the diamond business was a part of her. Just before he died, her father asked her to help him in the business he'd started in Italy twenty years prior to moving to New York. Working with him, she discovered aspects of her father that she hadn't been aware of, and she was grateful to have known him this way. She was now continuing the business.

Like some of the other women, Hannah, now in her fifties, came from a long line of diamond cutters. They had started in Poland in the early nineteenth century, immigrated to Amsterdam in 1865, then to Antwerp around the turn of the century, then to the United States in 1939. She encountered resistance because she was a girl, however. Her grandfather and father had sights with De Beers. Her brothers apprenticed at the wheel, but hated it and went into other professions. She remembered going to her grandfather's office after school and watching the men sort. She'd ask, "Why can't I go into the diamond business?" and was told, "Girls don't go into the diamond business." They laughed at her, yet she'd sit in the office, thirteen or fourteen years old, and her grandfather would show her some polished goods, ask her opinion about which diamonds had the better color, and she'd answer. She could tell such distinctions early on. But her mother wanted her to become a doctor, and daughters in Antwerp who went into the business were only sorters "shoveling *briefkes*," a Flemish put-down meaning rotating the parcel papers. Still, no one ever said a blanket "no" to her.

"It's not just 'no'; it's like a little kid saying, 'I'm going to become the president of the U.S.,' and everyone's laughing, 'Isn't that cute? Very funny,'" she said. "And if that person grows and starts going in that direction? They thought it was funny that I could tell color, and that when they were telling my brothers, I was there listening to the conversation. And it was all in Yiddish, and I picked it up on the street more so than at home. My father and I were very close. [Somehow I knew] it was OK to ask those questions.

"I was being prepped, but [my father] didn't even realize it."

◆

Michel said he liked being independent within the family—not a contradiction, he insisted. Making his own way within the confines and comfort of the family was really neat, he told me.

Becoming independent in one's family's business might seem paradoxical, yet the ability to make confident decisions within a collaborative and respectful family relationship was considered of symbolic and practical importance. It isn't easy, however. Many firms on 47th Street contained fathers in their eighties and sons in their fifties and sixties, which one dealer believed made for difficulties. When he was in his thirties, his father had voluntarily taken a back seat in their business. But this seventy-year-old dealer was still in charge, and he knew that was sometimes difficult for his son. Sometimes the fathers remained partners or sold the businesses to their children and then worked for them. Like the diamond dealer who was amused that her father bossed her around, others arrived at some rapprochement among themselves.

Extra tension derives from having to prove oneself continually. A person competes for business with other firms, with the other members of one's own firm, and finally, with oneself. While it is necessary to perform well in any business, an individual might be under increased pressure to succeed in these family businesses. One has to emerge from a childhood identity and position in the family to become a true and competent adult in the context of the family business. A dealer told me that he pushes for his opinion on a business decision all the while apprehensive that his way might turn out wrong. Ego is on the line; you don't want your brother, uncle, or cousin telling you that you made a mistake. The decisions are constant, and the tests unrelenting. Always on guard against making a mistake and ever vigilant for a slim opportunity, a person is on edge because of second-guessing, reconsidering, and fearing mistakes. Anxiety is prevalent, and slim profit margins are unforgiving. And if you make a mistake, someone might relate this fault to an ancient childhood trait or to family history in some uncomfortable way.

Starting out is difficult. A man in his fifties, Steve, recalled how insecure he felt in the beginning, when he was nineteen. He sat at the wheel for six years, cutting, afraid to make a mistake. As he gained confidence, his relationship with his father improved. The father, now in his eighties, was the senior partner, so his decision ruled. They respected each other's opinions, basically shared the same philosophy about buying and selling, and overall agreed about how to proceed. But tension continued. Just that morning, they'd had a little "joke."

His father told him to do something minor, and Steve said: You're still directing me; how long will it be 'til you stop that? Laughing, his father said he wouldn't be able to direct later on—so he might as well do it now. Steve said that so-called joke would have rubbed him the wrong way years ago. He was very sensitive then, and his father was difficult, a larger-than-life man in charge. Their interpersonal problems accompanied the business. As he wrested more authority with growing confidence, his father let him. When Steve became a father, things improved. He empathized with his father and felt more connected.

Paul, a manufacturer around seventy years old, remembered his internal conflict early on. Unlike those in his father's generation, people his age did not automatically go into their fathers' businesses. Born in Europe, he moved to the United States when he was very young. He wasn't sure what to do and ended up joining his father.

He and his brothers started practicing by shaping potatoes when they were children, and they joined the business at age seventeen. He called his need for independence from his father a form of wrestling. His father was critical, and he had to prove himself. Paul's father believed that you had to buy from the worst man on 47th Street and get the worst stone. His father used to criticize his enthusiasm over a particular stone. His father would ask: Why do you want to buy it? Just because it's pear-shaped? No? A marquise? *No?* The only right answer: to make a profit. And when he wasn't sure of himself, his father criticized that too. He had to learn to make a commitment and to know his emotions, to not be too eager, to understand his fear, and to feel what was at stake.

Worming her way into the business was Hannah's act of independence because her family had constructed obstacles when she wanted to join the firm. People outside the family urged her father to consider her before he was willing to take her seriously. Her father cited problems—claimed the bank wouldn't deal with her, for instance. Though no one in her family saw her talents and resolve, she believed she was being prepped. They answered her questions. She imbibed the diamond lore, the Yiddish, the characters of diamond industry heritage. She progressed in the business gradually. Then after she was allowed to buy and sell rings, they resisted her becoming a manufacturer.

Her father would say: It's not so easy; we have to find a very good place for you to cut. "What's not easy?" she would reply. "All we have to do is pay for my apprenticeship and you rent a wheel from somebody, and he pays for me to learn." "No. No. We have to think about it. I have to find someone." Once when she and her father were at the club, however, she asked her father who was a good cutter. He pointed someone out to her across the room. So Hannah got up and asked this man, "My father says you're a pretty good cutter. Would you teach me to cut diamonds?" When she turned around, her father had disappeared, gone back to the office, embarrassed.

The man said sure, he had an extra wheel. "Let your father give you some stones, and let's make the arrangement," he said. She pushed for each accommodation that followed. Her father agreed but didn't want anyone to know that she was in the factory. But she enjoyed the work and learned how to block a stone (give the first shape to the diamond after it's been sawed or cleaved). Then she would run upstairs and help her father sell rough so she could learn that aspect of the business. When she wanted to go to Antwerp to buy rough for her father to sell, he didn't want her to go; when she insisted, he sent her at her own expense. She made contacts. Her mother helped. In one case, her mother called a buyer because the buyer's wife didn't want the young woman calling her husband. Then Hannah would take the call and continue the deal.

Most of the stories I compiled about independence enshrined a symbolic turning point in which the child learned to be responsible. Entrusted with decision-making from the start terrified the new recruits but launched them onto independent thinking. They had to buy and sell on their own. For example, Atul's father gave his son $10,000 with which to buy diamonds however he saw fit. When he'd bought, his father told him to sell. Atul was really on his own, and learning occurred quickly. Although Steve felt he could discuss his uncertainties and was guided gradually, he thought his uncertainties persisted longer because he was not allowed as much leeway to make his own decisions. He seemed wistful about not having experienced Atul's sink-or-swim initiation.

Herman's father behaved like Atul's father. Confident today, Herman wore an expensive tailored suit and sat behind his shiny wooden desk as he recalled those early days. When Herman surprised the family and joined the business thirty years ago, he brought different ideas and battled constantly with his father. Finally, out of frustration, his father said, "Here is a sum of money. That's your stake. Leave. Do whatever you want. You can draw my inventory if you need to. You're on your own." Herman understood that he wasn't really on his own. His father had taught him how to sort, all about prices, being a polisher. Herman did things his own way, and though his father disapproved, it worked. After about six months he sat down with his father and said, "How about we're equals?" His father loved it. Herman had proved himself.

Herman backed up. He had been impatient with the kind of selling his father had been doing, and he was eager to try a new way: using direct mail. The mailings to solicit business replaced sending salesmen out for the same reason. After he sent out the mailings, he would call the recipients. This wasn't the way things were done. As Herman recalled, his father sputtered, "What kind of diamonds do you sell on the phone? What are you, crazy? You going to sell diamonds to people you don't know on the phone? You're going to send letters to people you don't know, and they're supposed to come to you? Why would they do such a thing?" And Herman answered, "Why not?"

Herman wrote letters to businesses describing his unique source of diamonds. In those days salesmen traveled. They'd come to New York, sit in offices, get merchandise, then go out into the country to try to sell it. Relishing his recollections, Herman recounted that his mailings said that though it's nice that salesmen sit across the desk from you and have a cup of coffee, and you talk about each other's children and grandchildren and so forth, he proposed a different way.

"'How about I save you five or six or eight or ten percent—and we won't talk baseball!'" he said he wrote in his letter. "'You're just going to buy diamonds from me. I'm going to deliver the best I know how to deliver for the price because I want your account—and we're not going to be friends.' And it worked! People actually . . . I got responses! You could have knocked my father over with a feather! He used to walk around saying Americans are crazy. 'What is this, you do business with someone you never heard of?' My answer to him was: 'I'm sending them goods. If they like it, they'll send me their money. If not, they'll send me back the goods. Where's the risk on their part?'"

Herman was redefining and reframing the business possibilities. He was "hungry" for the new method to work, and it did. The hunger was the necessary ingredient for business. He did not see this hunger in his own children. It was the trait Jim called "fire" that was also lacking in his son. Herman had a vision and went for it. He ran what he termed "crazy" full-page ads of odd-looking houses captioned, "A diamond house with a fresh perspective. Curious? Call." He got hundreds of phone calls and more than two hundred accounts. His father thought he was nuts. Herman's satisfaction in retelling the story was a satisfaction borne of deep affection and respect that the two men enjoyed with each other. He admired his father more than anyone else.

His father had taught him how to buy and sell. For example, his father was on vacation when a large order came in, and there weren't enough goods to fill it. Herman called his father, who said, "What do you want me to do? Come home?" "Well, what do you suggest?" asked Herman. "What do I suggest? Buy the goods and fill the order!" He did as he was told, and he called his father to tell him.

"And I said, 'I bought the goods. I filled the order.' 'What did you buy? Who did you buy them from?' All kinds of questions he asked me. He said, 'OK.' He hung up. My mother's description of what transpired on that end [of the phone], which I couldn't see, was that he almost fell out of the phone booth laughing and said to her, 'Boy, did my son buy a *strop*! Did he get stuck!' "

Herman paused to make sure I knew what a *strop* was. "If you buy the Brooklyn Bridge, you bought a *strop*. If you buy something worth $5 for $8, that's a *strop*." So Herman's father told his wife what Herman had bought and what he'd paid for it, adding, "Achhh, did they take him!"

Then, when Herman's father came back from vacation, "He came in, sat down and looked at what was left of the merchandise and said, 'Tell me, do you think you bought it well?' And I said, 'Sure.' And he said, 'Looks good to you, right?' I said, 'Yeah, looks wonderful.' He said, 'OK, why don't you just go out and sell it and don't come back 'til you sold it.' What I did, I wandered around for about ten days, and I realized that I really had been taken very nicely, came back, and said, 'I don't think I can sell this stuff.' And he said, 'Long as you learned something, that's OK.' That was his attitude. That's why I admired him. I don't think I ever had that kind of patience. He understood that was the only way to learn. I really admired the guy. He had the capacity to do this."

And Herman leaned back in his chair, stretched and sighed, savoring and missing his father.

◆

Some loved the business and enjoyed the pace of the work. Some thrived on the lack of predictability, the excitement of not knowing what would happen in the coming day. They liked the pleasure of a good deal, a satisfying sale, a buying coup, the challenge of being honest, cultivating a good name, enshrining a reputation, continuing a family legacy. They enjoyed the people they'd known for a long time and the new people they met. Many enjoyed the travel, others the satisfaction of working for a firm that defined their family nestled in the safety of the mostly Jewish enclave—helping it grow, ensuring continuity. It fit, even when it chafed.

At the same time, people pondered whether this was a suitable business for their children. They witnessed bankruptcies. They knew the people who failed, those who were resurrected, those who could not pay for goods. They knew the anxiety. Enduring the hard times was painful, and hard times were unavoidable. They knew how fragile their prospects were and wondered whether this business made sense for another generation.

Some quietly endured the family business. It had been the default fallback, all choices looking equally tepid. Family expectations exerted pressure. The wish not to disappoint was coupled with an unwillingness, perhaps a fear, to fight for something different.

A middle-aged diamond wholesale dealer and a jeweler entered their family businesses after stints in the Army. When I asked both men about their decisions to enter the business, they looked empathically at each other, and each held a long breath. Then they said there had been definite ups and downs. They didn't want to say more.

Some revealed their less-than-enthusiastic attitude by telling me they did not encourage their children to follow them. The economy was tougher, people weren't wearing ostentatious jewelry, New York was not the center anymore, they said. They distrusted the times. Were they also disappointed with their own lack of enthusiasm about work? Did they regret not trying something else?

One middle-aged broker, wearing an embroidered *yarmelke* slightly askew, bent his body almost sideways when he complained to me. He didn't like the influx of Hasidim here. He didn't like how the Indians who specialized in the smallest diamonds had automated the business. He didn't like these single-faceted diamonds in the tennis and solitaire necklaces sold for $100 or $150. He sarcastically mimicked the seventeen-year-old who looks at her parents adoringly after they buy one for her—the ultimate sacrifice! Cruddy jewelry, he said; if only they knew. That's the way the business was going.

He continued bitterly that he didn't want his sons to go into the business. Business was too hard and too slow. Markups were too small to make a living. His sons make more than he does. This is just a *social* club, he said, waving toward the small nuggets of men hunched at the tables. Look at them: people here all the time, playing cards. They come in late. It's a lazy man's business. He pointed to a dealer about his age, saying: He went into business because he couldn't do anything else. He'd probably like to get out of it, too.

Some of the most successful men who liked the business had grave reservations about their children following them. Some had told their children to go into something else. They bragged about the fields their children had entered.

A ninety-two-year-old Antwerp-born dealer had a contrarian view. Sturdy, with fleshy face and alert eyes, he made his cane look like a snazzy accessory. He looked the prosperous burgher, set in deep bourgeoisie. It's a good time to go into the business, he trumpeted serenely. Why? Because you should know that business is tough, and that it has cycles. That way, you get properly prepared. It was better to begin at a difficult time, like the mid- to late 1990s, to dispel unduly hopeful expectations and to be effective in the real business world, he said. If you started at a prosperous time, you would have unrealistically high expectations. A dealer thirty years his junior agreed completely. Business was always hard.

And daughters? As in other families, some daughters were invited and some were impeded by their fathers. Three young Orthodox men thought their sisters would not be "allowed" to go into the business—though many Hasidic young men were entering the business. Some made the distinction between the retail and wholesale sides of the business as well as between wholesale trading and manufacturing. Several fathers and mothers told me their daughters had gone into retail jewelry sales, where being a woman was an asset. One trader said he would hesitate to encourage his daughters to enter the wholesale business, but he thought he would feel the same way if he had sons; he simply didn't have enough confidence in the business now.

A woman's granddaughter had wanted to go into the business, but her diamond dealer son objected. "I think Sonia wanted to go into it, and Allen wasn't interested," she said. "I knew about it. They were very open about it. Sonia said to her father, 'You don't want me to go into the business.' And he said, 'No. I don't think it's a business for a woman.' I imagine he must have given a reason to Sonia. It is hard. It's a man's world, and they're not very gentle. And also, socially, I guess he doesn't feel she should get married to a diamond man. I have a feeling . . . he felt she should have a more intellectual [interest] . . . although you can find an intellectual businessman too."

Another trader intended for his firm to end with him. Diamond traders are peddlers, he told me with scorn, not a field to which one should aspire. But a successful Israeli diamond trader hoped his son or daughter would join his business after trying other fields and studying Asian languages, which would serve them in good stead with whatever course they chose. And a manufacturer in his seventies commented sheepishly that his daughter had wanted to go into the business years before. Though his sons had tried it and left, he had discouraged his daughter on more than one occasion, and now he regretted it.

◆

What does being a woman have to do with the business? At the least they listened to the stories that the men brought home. I became aware of some of the involvement when talking with a diamond dealer and his wife about the business. As the diamond trader spoke about this topic or that, his wife offered a running commentary—sometimes fairly acerbic—about the people involved. One man was a "blowhard," another "unappetizing." The wife of a man seeking DDC office, she reported, was unhappy he was running this time.

At another gathering, a family was talking about the day's events at dinner. The daughter-in-law asked her husband: If a stone is low-graded—say, an XYZ—isn't it a fancy? No, her husband answered, it's just a yellow. She then asked: What percentage below Rapaport list is a diamond discounted? Diamond details, stories, problems, and personalities are no doubt dissected around many dinner tables as people "have diamonds for supper."

But women also work in the trade. During most of the twentieth century diamond exchanges accepted women only as widows and daughters of deceased husbands or fathers. Even then, their membership status was inferior to that of men. Women were usually not allowed to vote, for example. In 1919, a woman applied for membership to the Amsterdam diamond exchange. She was refused but then was accepted after fervent debate (Lipschitz 1990). In 1952, Jannie Roselaar became the first female member of the Amsterdam Diamond Exchange. Her story sheds light on women's experiences at that time:

"I learned the trade as a girl; this was before I was married, of course. That was in 1911. I can remember them building the Exchange. When I learned to cut, it was still done by hand; later I learned to use the machine. After we married, my husband and I used to sit and work at the same table. He was a very good cleaver, and I always paid attention so that I picked up quite a lot about cleaving from him. Later on I was no longer allowed to work, which was a great pity. Men used to think that married women in Holland never did any work. But women used to work like slaves. They only got to bed at two in the morning, when all the household chores had been done, and they had to be up again at the crack of dawn. Men used to ask each other what their wives did all day. 'Nothing,' they'd say. It was only work if it was away from home.

"When my children had grown up and no longer needed me to look after them, I asked my husband if I could work at the office. But he said it would be too embarrassing for him! He just couldn't do it. I could do whatever I wanted; I was completely free—even more so than other women. But office work? Out of the question. One just couldn't do that in those days. His position was too important. On the Exchange, they would have said, 'Have you heard? Roselaar lets his wife go out to work!' That's why there were hardly any women at the Exchange.

"After the war I was completely on my own. I could have let the business go, but I decided immediately, 'No, first give it a try.' I had learned the trade well, but I had been out of practice for more than thirty years, and that's a long time. Even so, it went well. I worked on the Exchange for a number of years, when it was still frowned on. Eventually I became a member, and that was quite a struggle. The procedure dragged on for one whole year. The foreign exchanges also had to give their permission, although they made less of a fuss than the people here did. I still remember clearly how one member—long dead now—actually resigned from the various committees in protest against the admission of women—even though we were friends. I was very careful. I was never pushy. Slowly but surely I got a place—and what a place! I couldn't have wished for better. In 1952 I was admitted as a member. Another couple of years were to pass before more women were allowed to join, but they had a rather easier ride than I had."

Female brokers weren't allowed onto the Exchange floor, she continued. "They let the stingiest *chazzers* [pigs] in but not one woman. Once when I was in the hall there was a woman at the foot of the stairs . . . a real lady. But even she wasn't allowed upstairs onto the Exchange. It was unbelievable!" (1990:122–25,133).

Change was inevitable, however. The New York DDC bylaws were revised in 1980 to conform with state and federal sex-discrimination laws. Until shortly before then, the only female members had been widows of dealers. Manufacturer Ethel Blitz publicly described her long route to becoming the first full-fledged female member of the DDC in 1978 at age thirty-one (Friedman 1979). Six months earlier she had become the first woman allowed to buy from the CSO, eager to encourage more manufacturers at the time. But though she traded in Antwerp, she was barred from its bourse. Permitted to use the ancillary facilities at the DDC in New York, she was not allowed to show goods to customers. When she was admitted to membership in the DDC, twelve widow members were promoted to full membership. As Friedman reported, the Hasidim were worried that approval of her membership would trigger a cascade of female membership applications, but stiff membership and financing requirements for admission as well as the difficulty of the business have kept the numbers low. When Ethel Blitz wanted DDC officials to change the "Dear Sir" that headed her letter of acceptance, they reportedly told her she was being too sensitive.

Nicole Polak, like Ethel Blitz, came from a long Antwerp line in the diamond business. Though both women had to adapt to restrictions imposed by Hasidism, she had an easier time and told the same reporter, "I can't say I'm a pioneer. I feel more like a princess. I like the fast pace of the business, the feeling you're on Wall Street." Penetrating the industrial diamond market was easier than the gem world because it was freer from restrictive rules about women, she said.

About fifteen years later, Bates (1995) profiled Ann Rehs, Edith Lipiner, and Norma Haas, who said they were successful in a male-dominated business because they worked hard and were therefore taken seriously. Some of them felt advantaged as women and noted that women in the retail trade like dealing with women in the wholesale trade. Carol Weiss Enlander, profiled by Furman (1996), was a determined daughter of Holocaust survivors with no connections to the diamond business who became a member of the DDC and served on the board of directors.

◆

What did the men think of the women traders?

"There are a few women coming into it, especially in the retail part of the thing," a manufacturer noted. "Quite a few women now. Manufacturing— not many. There are a few women cutters. In the war, when men were in the Army, we trained women to cut diamonds to replace [the men], but once they get married. . . . Hasidic women have too many children to do this. Or now they work on computers at home. But in the diamond business manufacturing—very few. And in manufacturing, don't forget, it has a lot to do with being Jewish. The first thing a Jewish girl thinks is of marrying a Jewish doctor and therefore the idea of entering the diamond business enters very few heads to begin with."

Some men I spoke with in 1986 were pleased that a woman was running for arbitrator. It was time for a woman to make it, they decided, and they thought she would win because she was well liked. Though she lost by a small number of votes, two other women ran successfully for arbitrator a few years later and at least one attributed her win to support by the Hasidim.

A trade writer's book discussed some of the trials the women in the business had to endure, including the lack of women's bathrooms in the club until the mid-1980s, when the club was renovated. He listed the negative beliefs: women should not do this kind of business, are not good at the business, and are easy to bargain with. They had to endure condescending remarks that husbands "allowed" them to work, and they had practical difficulties of working with Hasidim, who are not permitted to look at or touch women other than their wives. But pragmatism was a good antidote. As the first woman CSO broker noted, "The prejudice and the 'old boy' network are disappearing, and attitudes toward women improving. At the end of the day, I believe it's the competence that counts" (Shor 1993:206).

A man in his eighties thought that women's impact in the wholesale trade remained minimal. However, strong women had been in the trade in the past as well as now, he noted. Their strength had to do with their personalities and with the fact that they had handled large, important stones over the last decades—not with their gender.

A middle-aged trader believed that women were accepted much more readily now and cited the two female arbitrators. On the other hand, were they respected? Not by the Hasidim, he felt. Yet some Hasidim must have voted for the women who had just been elected arbitrators, he said. And there were many women in the retail trade. He himself saw no difference in dealing with them. Another middle-aged trader wondered why there weren't more women in the business.

In fact, he said, women are just plain more attractive to deal with—more charming, more intuitively in touch with the romance of the stone. Look at that woman sitting next to her Hasidic son, he said, gesturing. Her son had urged her to enter the business after her husband died so "she wouldn't go crazy." She had been in a business before, so she was good at it. This is a perfect business for women because they can work a few hours a day while their children are in school and supplement their husbands' income, he said.

A DDC board member was proud that the New York bourse had been the first in the world to change its bylaws to accept women as full members, in 1980. In 1982, the WFDB followed. The Israelis objected at the time because they feared that women would have an unfair advantage—their fabled "natural" predilection for the gems. The DDC official thought many women have gone into the Israeli trade because:

One—New York was proportionately more dominated by the Hasidim than Israel, and Israel was less prudish in general. In New York's DDC women had to wear skirts and dresses. There were fewer restrictions in Israel's club. He heard someone in Israel object to a woman's short skirt, saying her *pupik* [navel] was showing. Here they would have killed her for a skirt that short—not really, ha ha, he added hastily. Two—Israeli women were more aggressive in general, possibly because of the universal army service in which women participate for two years at age eighteen. And three—the better security in the Ramat Gan diamond complex assured safety. The four interconnected diamond trade buildings had a terrific security system. No one has to go to the street to do business, unlike in New York.

However, an Israeli dealer I spoke with did not see much difference in women's situation in Israel versus in the United States. In both countries there were few important female dealers. He could think of only one significant woman dealer in Israel.

Other men offered their opinions. An elderly manufacturer, proud of his daughter's jewelry business and factory, noted that the main buyers at retail stores such as Neiman Marcus, Van Cleef & Arpels, and Cartier are female. He insisted that women have made great strides, as did another manufacturer, whose daughter had joined him in the business. He realized it was hard for women to be taken seriously. Marcel, a successful dealer in his late fifties, said changes in the status of women in the diamond business were phenomenal, and opportunities were burgeoning. He pointed to the hard work and success of his efficient, brainy, and polished wife.

A banker on 47th Street whose branch was almost entirely devoted to the diamond industry claimed completely even treatment for his female and male customers. He speculated that his bank had accounts with twenty to twenty-five businesses in which women were the dominant figures. We do lend to them, he said. We do give them credit facilities, he added, perhaps rhetorically. And they are just as capable as men, maybe more so, he insisted. He said women were being accepted more, especially among the ultra-Orthodox. Few banks have an interest in diamonds, he said; they are a luxury item, and when there's any trouble, the diamond and jewelry industry is the first thing to get hit. Women are not having any harder a time than men, he said, referring to the Asian financial crisis of the mid- to late 1990s. Also, the supply from De Beers is very tight since they're controlling the marketplace. That wouldn't be harder for women as opposed to men, either, he said. In dealing with a female entrepreneur seeking credit, he said, we would interview her the same way we interview a male, based on financial strength.

◆

Some of the women I talked with in the mid-1980s were no longer in the business in the mid-1990s, while some had continued. Their views and experiences ranged widely. Some women interpreted what men said as condescending. For example, women didn't like that men noticed that some of them worked part-time. Were they implying a lack of seriousness?

One said: They don't question a man when he leaves the club, but *her* they question! It doesn't matter what time it is. If it's *erev* (the eve of a holiday) anything, they say, "Aha, it's *erev* whatever," meaning that *that's* why she's leaving early. Some of the men even question her presence in the club with remarks like, "What are you *really* doing here?"

Being a member of the club had made Lucille a feminist—in the worst way, she told me in 1986. Willowy with short black hair in a chic cut, Lucille was middle-aged and gave off an air of confidence and competence. She had helped her father, now worked alone. The unwanted attention she received from men—the jokes, the remarks—was a constant irritant, she said.

Because her husband had a separate diamond business, there were comments about that. Regarding an article that was written about women in the business several years later (Bates 1995), she thought the women had soft-pedaled the truth and put a positive spin on their experiences, perhaps to maintain whatever fragile position they had.

People assumed her husband was supporting her, she said. When she traveled on business with him, they assumed he was taking her along. Most people in the DDC haven't figured out that gender-wise, the world is different now, she said. She offered this example: after perusing goods in the club, men customarily get up, nod, and say "gentlemen," in parting. It isn't a big deal, but it bothers her.

Being on her own can be hard, she said. Before the era of cell phones, she faced one dilemma involving telephone calls at the club: when she was at the club and had to get up to answer a page and go to the phone, she took her entire briefcase, her inventory, with her. It was awkward when she had a customer looking over her goods at the table. She didn't want to forgo a call, but she also didn't want to leave the customer alone with the goods. There was anxiety either way. Also, taking time off is hard, she said. She'd like for the work to be less pressured, but that's impossible. It would be unprofessional to take time off, and she wouldn't be taken seriously if she did.

While we talked one time, a man came over to use her electronic scale. Wordlessly and without a sign to her, he put his diamonds on the scale and weighed them, then motioned to Lucille for the little shovel for scooping them up that was slightly out of reach. She raised her eyebrow at me while this was transpiring, as if to say, "See what I mean?"

Before Hannah became a member of the club, she was paged by her father's name. It is customary to be paged by both first and last names. So, when Hannah's mother requested Hannah at the club, the name "Myron Aronson" sounded through the loudspeakers. Her mother was exasperated and protested to the switchboard that she wasn't paging her husband. "I know who I'm married to. Why don't you just page my daughter?" The positive aspect of not being a member of the club was that Hannah did not have to pay dues. She and her father used to joke that if a problem occurred, she couldn't be called to arbitration since technically she was a nonentity. Later when she and a partner had their own business, their Hasidic factory workers talked to her Hasidic partner, not to her. And remarkably, at her DDC admissions interview to which she'd shown up in a business suit, they had asked her if she was going to wear a sundress to the club.

When club membership became available to women, and the widow members were elevated to full membership, some were upset because they then had to pay full membership fees. They didn't seem to understand the positive implications.

The countless tiny booths owned by separate individuals in the exchanges at the floor level on 47th Street are packed together sardine-like as private customers stream through, looking at the antique jewelry, diamonds, and other gemstones displayed. The exchanges provide good contrast to the upstairs offices and the DDC. The narrow corridors in the exchanges are lively, filled with people speaking with pressured animation. Ruth worked in one of the exchanges. She had been in her teens when she and her family fled from Europe during World War II and went to South America where she started to work in the business. When the factories moved to Cuba during the war, they needed cutters, and she became a girdler for a number of years. After the war, the companies left, and the industry was disbanded. She came to the United States, married, and worked in sales until she had children. When her children got older, she went back to work. She liked putting the pieces together, but not the designing part.

Ruth introduced me to Estelle, a woman in her thirties, holding forth frantically and flamboyantly in the exchange booth next to hers. Estelle told me she had majored in English and couldn't find work until she landed in a jewelry store. The store realized she had a facility for selling, and she started to deal her own goods. Then she worked at Sotheby's, made a lot of contacts, and opened the booth in this exchange with another woman as partner. She said she likes her independence and the people, despite some rough characters. There is nothing better in life than making a profit—what a high, she said. But you can't depend on it, she added, because the business goes up and down, and it can be really hard. The wife of a prestigious jeweler across the street told her she'd survived pogroms and the two world wars—but nothing was as bad as the 47th Street cutthroats.

We went outside the exchange and crossed the street to the Hebrew National restaurant counter, where she bragged of the many deals she made there. She embraced people she recognized as they passed through.

Back upstairs in the club a young woman of Italian ancestry, tailored and svelte, strode through purposefully, carrying an elegant briefcase with her goods packed neatly inside. As we talked in the club, she adorned her story of how she entered the business with details about what it's like being a woman in the field. She had shelved her plan of going into architecture in order to be in business with her father.

This work is so varied and full, she told me, like *life*. Business is like life—personal, direct, varied. It's absolutely *no* problem being a woman, she said; there is *no* difference. You throw the dog a bone and that's it. They'll deal with you, no matter what—whether you're black, a Turk, a woman. That's what's so amazing, she said. Maybe it's because these people are survivors that they're so pragmatic. Whatever—she loves it, and she won't go back to her other field.

A seventy-five-year-old woman recalled that she and her sister-in-law had worked for a short time in their husbands' office in the 1960s. They did some bookkeeping, answered the phone, and attached tags to rings that their shop produced. The work was boring, but she enjoyed connecting faces to the names she had heard about in the colorful stories her husband had brought home at dinner. She liked going to and from work with her husband, and he didn't expect much for supper. Men were so abrupt on the phone, however! When they found out her husband was not there, they would hang up. "'Is Herb there?' 'No, can I give him a message?' Boom. Really, it was rude."

One woman made jewelry and, through a distant relative downstairs in an exchange, developed contacts and sold some pieces for a small profit. She appreciated the personal and intimate nature of 47th Street.

A manufacturer in her forties said she believed that the key to making it as a woman in the business had to do primarily with reputation rather than with gender.

A wife of a diamond trader commented, "I knew it would be very hard [for women]. It wasn't done, I guess, or very rarely . . . [only] the widows." And her granddaughters, daughters of a second-generation dealer? "I don't see it," she answered. "It's completely strange to these girls. All they know is he brings home the money and that's it, you know. He took them into the office a couple of times and that's it. . . . My father was in the diamond business in Antwerp and here also. My father didn't talk about business. It was not made for women, see? He told stories but nothing else."

A female trader believed that there were more women in the business but that fewer of them were club members. The lack of a women's bathroom for years had been symbolically important. Women were still not taken seriously, she felt. Then again, she wondered, how serious *are* the other women?

As with her male and female colleagues, Lucille often was critical of other traders. Irrespective of gender and ethnicity, she was on the alert for traits in others she considered bad for business. She was quick to avoid those whom she identified as such. Like some, she found Israelis untrustworthy and unduly bold as a group: they'd taken too long to pay for the goods they'd bought from her and that was risky. She thought her sharp judgment had helped her avoid traders who did not seem financially secure.

One of the female traders described the difficulties of the marketplace in the late 1990s and concluded, "Men and women both take their licks; at least there's equality on that. Absolutely."

Iris had been a broker for years in another office when friends suggested she become a member of the club, something she hadn't considered. Many were going out of business at the time, but she became a member nonetheless. She kept her business small, still brokered, but owned her own goods as well. She didn't particularly trust the trade, and she was grateful each of her grown children had gone into other fields after trying the business for a while. She thought she had a special knack with the customers and could "bring them in."

Being a woman in the business was no more difficult than in any other trade, she said. What was hard was the anxiety of carrying other people's goods around—the constant worry, the high insurance, that kind of gauze-like tension that is indistinct and knots one's stomach.

A number of women hinted at sexual harassment and discrimination but would not elaborate. One, however, was eager to vent: "They get upset with women who want to fight back," she began. "You're just calling for credibility. The men haven't moved that much forward. Look, sexual harassment has never been discussed in the industry. We'd have more sexual harassment than any other industry because no one knows that that's not what's done or said. The men think sexual harassment means I make a pass at you, or I rape you. They don't know telling dirty jokes and ignoring is part of that. But policy manuals in all industries are being rewritten and they have to discuss protocol and how do you behave. [In this industry] this is never mentioned, and you won't find any of the women ever mentioning it. [Are the women being harassed?] I think so. And they're not discussing it. I've seen it. They're not in the factory but in the offices. . . . I think so. But do they discriminate against women? Yes, I think so. For instance, when my father died, I had people who turned around and said, 'Well, I know I owed your father some money. Your mother doesn't really need it, and you have a lot of money.' Would you say that to a son? I said, 'Excuse me, that's not the way it's going. You owe and that's it.' I said, 'The balance is still due.' He still won't talk to me. And they think the daughter doesn't know anything about the father's business, which is funny. He's claiming the same *shtick* [to other] widows who don't know their husbands' business."

Elaine batted the gender idea around with me. She started in the diamond business twenty years ago, working part-time, then became full-time years later. After working six days a week, she jokingly decided: I need a wife, an old-fashioned wife! In the beginning the phone calls would come in, and they'd always want to talk with her husband and business partner, Marcel. The secretary would pass the phone to Elaine. They would protest: I want to speak to Marcel! And she would explain that she could handle the business. If the men need something that I have, she said, then they're more amenable. It's been a man-dominated business for centuries, like Wall Street, and it's also an Orthodox community who by their own definition won't have anything to do with women, she noted, adding that she feels lucky to work with Marcel. The business is difficult for everyone, even with family backing, she said. And women still constitute only a small portion of the industry. Women are active in the selling but not in the market (47th Street), she said. I'm a fixture now, she added, but it took a long time to be taken seriously. And even then, she encounters sexism—comments like "She's here now but maybe she'll have a baby next year."

She did not think gender discrimination was much of a problem, however. Once they get used to you, she said, it doesn't make any difference. Like others, she distinguished between the retail and the wholesale trade. The retail customers have no hang-ups about women. In fact, being a woman is an advantage in retail, she insisted. It's a relief for the customer to not have to go through the men, the shenanigans, she said. She sells to retailers, wholesalers, and manufacturers, but not directly to individual consumers.

Despite the obstacles for women, she said she'd love for her daughter to go into the business. A woman is less likely to come in from the bottom but would still have to prove herself on 47th Street by dealing with comments like "What can I do for you, my little girl?" "Can I have a D flawless on memo?" "My dear girl, I should give it to you? Why would you want that?" If her son went into the business, she said, he would be treated more respectfully. The men in the business coddle their women, she said. They can't figure out why women would go into the business. But I absolutely have no problem with it anymore, she said.

Then she reflected and added: I must say I'm much more careful because if I slip up, they'll say, 'What do you expect?' That's the way I'm thinking, but I don't know if that's what would happen, she said. Then she added: But it's probably no different in other businesses for women. One challenge that women in the business face, she said, is how to socialize with other women in such a male-dominated industry.

Most of the women had little time for socializing, however. Some of them told me they enjoyed friendships with other women in the business; others seemed to have little to do with them. Some of the women derived reassurance, empathy, and at times real help from other women. Lucille told me she helped Sue get started by referring customers to her and by providing friendly support. Newcomer Esther was grateful to Elaine for steering her to the Women's Jewelry Association, which held meetings and promoted networking. One liked the camaraderie and empathy among the women, but felt isolated because she was on the go a lot.

Esther had recently opened her one-person business in New York and was creating the New York branch of her father's European family business. She said 95 percent of her work involved dealing with men, but she thought being a woman was an advantage. As other women who spoke to me, she considered her positive experience an exception. I think men think it's nice, she explained. We call it being among the sharks here in New York, but I find that people are very helpful, she added. It all boils down to the individual, she said, and that's why it works for her.

◆

Lucille at times criticized some of the Hasidim—they ran around on their wives, had many children but no family life, pinched women on the behinds, right there in the club—but she had little problem working with them. She used some of her father's Hasidic contacts, and with time they became used to her. If you have the goods, she said, then they want to see them, and that's the bottom line. She puts the goods on the table, they pick them up, and there is no contact. Very respectful and fine. "Listen," she said, "people would do business with a monkey if the monkey had the goods." She paused, considered, then added with a short, sharp laugh, "Maybe sooner with a monkey than a woman. . . ."

She usually ate her lunch at the club each day with the Hasidim after they were done with their noon prayers. Some of the men in that room knew she was uncomfortable, and they usually made room for her. If someone felt she was sitting too close, he moved away.

A woman who had been in the business for about five years believed that women were respected at New York's DDC, but you can't talk to the religious people, she said. One time she needed to sit down and she accidentally tapped someone on the shoulder to ask him to move over, and he nearly had a stroke, she said. Everyone else was explaining to her, excusing him, "He's a religious man, he's a religious man. . . ."

Another woman, Iris, found the Hasidim puzzling. "They're not supposed to interact with women, right?" she said, adding, "Well, they're phonies." She said she concluded a deal with a Hasid and was so happy about it that she went over to him exuberantly and said, "*Mazal!*" with her hand outstretched. He refused to shake it, and the others around the table laughed nervously. She realized too late that it was her mistake. Later, in a less public place, the man apologized for not shaking her hand and tried to shake it then. She refused and told him she made lots of money off that stone. He retorted, "Enough for *challah?*"—belittling her accomplishment, she thought. So she said, "Oh, way more than that!" At that point, he again attempted to shake her hand, but she refused, saying that he couldn't do it now after embarrassing her before.

Though she admitted her faux pas, she gave him no credit for wanting to shake her hand out of public view. Their subsequent overtures and retorts were a clumsy negotiation in a world of evolving roles and attitudes.

Akiko, a Japanese buyer in her late twenties, was employed by Takeda, a Japanese dealer, and wore a tag that said "Buyer." Akiko and I spoke about her four years of work in the ladies' room, the female safety zone of many male-dominated work environments. She told me she couldn't speak near her boss. It's hard to be a woman in this business, she vented. It's hard to be taken seriously, but she hopes to own her own business in the future. Single, she gets harassed at times by the Israelis, but, no thanks—she wouldn't elaborate. A couple of the Hasidim have asked her to go out with them, she reported, incredulous. Though sympathetic to them—they were born into this confining life, it's not their fault, and they must not disappoint their families—she disdains their lack of education. Many of the Hasidim never even went to high school, she said, and she has to simplify her talk when she speaks to them. In all, however, she likes the business.

Elaine was firmly practical about the religious men. Speaking specifically about dealing with the Orthodox, she echoed some of the others: There is absolutely *no* difficulty dealing with the Orthodox. It took them a while to get used to her. The younger ones are more reconciled to women being in the business—if you have the goods, she said. That's the important thing. In general, none of them will shake your hand, call you by your first name, or look you in the eyes. You won't have any problems with the young [Orthodox] people, she said. Most of the Hasidim are old-timers, however. On the other hand, they're very nice; they can be very personal. If I encounter any difficulties, she said, I pass them on to my husband.

Esther, likewise, had a practical attitude about the Hasidim. She said she even manages to work with Orthodox people, though not with all of them. "I speak fluently, Yiddish," she said. Religion is one thing, however; business is another. If I have the merchandise that one of them is looking for, if he doesn't want to deal with me directly and I need to give the merchandise to someone else, it'll cost more, she said. The middleman is owed a commission. So some of them will deal with me directly, she said. Some of them strictly will not deal with a woman. But with some of them I really have no problem, she added. I wear a jacket, so I'm not sleeveless. I don't want to intimidate them. It's the merchandise that needs to sell. The language helps a lot.

Joyce felt that the impediment to working with the Hasidim was the fact that she was in a solo business. I don't deal much with Hasidic men, she said, because at times they will not come to my office because I'm all alone, and they can't do that. So that's a difficulty.

One manufacturer said it was impossible for women to manage a factory with Hasidim.

"Maybe marking stones," he allowed. "There are a few women who mark stones. Not here. Maybe one. In Belgium if there's a man in charge of the factory and he's a Hasid, there will not be a woman in the shop, because it's not allowed. So that eliminates a lot. Religion has a lot to do with it."

How women, as well as men, reacted to such challenges perhaps reflects more on individual personality differences and capabilities than on gender. Individuals came to the business under unique circumstances. They were differentially motivated, had various goals, and came from a wide gamut of family backgrounds. Some were overcome by the obstacles; others saw the obstacles as unique challenges that lent them an ultimate advantage. If the diamond business is not yet a field for women, women are making it their own in their own ways, nonetheless.

◆

"Perhaps we spend the greater part of our lives fighting against the clock, trying to deny our limitations and the physical process of our aging," Cole has written (1992:250). "Finally, at the end of life we may accept reality and go with the clock." As elderly diamond dealers attempt to work, persisting through the limitations that aging brings, they seem to be living their preference.

Aging is not for sissies, as the expression goes. The diamond business is hard, New York City is hard, and staying physically and mentally alert is hard. Though the work of diamond dealing is primarily mental and social, most are also active physically, and the constant walking—from office to club, down and up stairs, in and out of buildings and offices throughout the district spread over several blocks—probably helps maintain fitness.

If he or she remains in fairly good health, a diamond trader can continue a way of life that supports socially as well as financially, provides a structure to daily life, and validates a lifetime work and personal identity. Despite the anxiety and tension of the trade, the stress of speeded-up electronic connectedness, and the harsher conditions of the business in general, old people press on with their work and derive satisfaction from its continuation. The diamond business accommodates them. The social cushion of the club seems adaptive to aging. Working means continuity for these individuals, but they do not idealize it. The hard realities of age—the problems and stresses that age brings—are discussed, lightly sometimes, less so at other times. Though leavened with humor and sympathy, judgment is never suspended. The past haunts as well as enshrines one's reputation.

Of course, people also retire—some voluntarily, others not so. Generally, a person stops work because of health or financial difficulties; sometimes he or she has been relegated to an undesired peripheral role in the business. Such retirement is dreaded. No one wants bankruptcy, frailty, or dementia, and no one wants to feel pressured to leave. A ninety-year-old man sadly told me that his ninety-two-year-old partner had died a few years prior. Though they used to run a big operation, he was only about "5 percent active" compared with then. His semiretirement was "kind of forced," he allowed. This is my office right here, he told me, and he looked little against his corner of the table in the club where he was sitting. If I come in a little late and someone else is sitting here, he said, I find another place for my office. Arnold was phasing himself out.

For some tasks, such as evaluating stones and assessing market conditions, advanced age and years of experience can be a definite advantage, given good health for the task. The judgment that these traders have cultivated over the years helps to tamp down panic and anxiety, works to modulate the unbridled enthusiasm of those much younger, and assists in understanding the vagaries of the market. The mix of ages in the DDC displays and tosses perspectives of heterogeneous origins, perspectives that complement and compete, together creating a more balanced understanding of the market, men, women, and mortality.

Many—perhaps most—of the people who did business with one another had known each other for decades. They were mirrors, compasses, and barometers for each other. They watched one another as if to check on themselves, to ask: How am I doing? Not so bad, perhaps. Age was the common predicament claiming them, but in the meantime they had one another to share with and to commiserate with. Insults were binding and could sometimes get better with age, as this story from Antwerp illustrates.

"One of the strongest chess players in the club has a rather sarcastic tongue," related a trader. "He's retired now and he comes to the bourse every day—he has nothing else to do. What he loves to do is entrap players. That's the great passion of his life. One day I saw him playing. This has to be told in Yiddish. . . . One day he's playing and a friend of his approaches and says, 'You've lost a horse (a knight).' 'Yes,' he answers, 'I'm only playing with one horse.' Meaning the horse is his opponent. But in Yiddish it has more connotation of stupidity, of contempt. Playing with a horse means you're playing with a real horse's ass. This guy is superior to most of those who kibitz and sometimes a kibitzer will give a piece of advice to one of his opponents and he'll say, in Yiddish, 'Why are you kibitzing? He can lose without your help, also.'"

As part of the same scene, the old traders were models for the young. The young traders were noticing them, planning how they wanted to be different or similar from them, and preparing for their own futures. They were seeing the diversity up close. Old age was not a theoretical, distant concept but was being woven into their daily lives. The younger ones were witnessing the older ones trudging through age, and they knew they were on the same conveyor belt. This realization lessened the tendency to stereotype the elderly.

The physical structure of the club also supported the social aspect. The open hall, the banks of long tables with chairs arrayed along each side so that people could talk across the table from one another, the chairs that could be clustered for larger impromptu gatherings—all contributed to the social dynamics of the club. The *shul* offered a place for prayer and study for the religious. The smoking room in the back of the DDC and the tables where older men often played cards and read newspapers also lent a convivial air to the surroundings. The luncheonette and adjacent restaurant, busy with those who talked about business or gathered with guests for a break, served as a release from the work.

Whether in the club or in the offices, business is carried out via social relationships. Through countless connections, people ask one another if someone has certain goods, when to expect payment, whether a third person is financially solid. This buzz is the medium through which business is conducted. News of personal health and daily life is woven into the workday. People share information about recent births, upcoming bar mitzvahs and weddings, deaths, illnesses. Shalom, the father of eight, was telling about his son's upcoming wedding in London.

Black-bearded, his handsome face sparkled with the telling. Trying to appear blasé, he enjoyed the flutter of the news. Heschel glimpsed him, offered *mazel tov* on the new grandchild, and Shalom laughed. They've only just announced the engagement! Shalom didn't want to bring his two-year-old, Mikhal, because it was too difficult to travel with her. You *have* to take her, protested one, while others shook their heads, agreeing that the logistics would be too hard. As they argued, one left to answer a page he had just heard over the loudspeaker.

Deaths were discussed in often subdued, matter-of-fact tones. Illnesses, the feasibility of certain medical procedures, the pros and cons of different doctors, and the promises of alternative therapies were discussed, fine points minced, predictions projected, and decisions shared.

One example illustrates such encounters: Ellis asks Walter why he is back at work on this frigid day so soon after his hernia operation. Walter brags that in about ten years—when he's ninety-six—he'll start to make concessions for his age. Walter whispers to me that the man next to us is ninety, and the one at the next table is ninety-seven. He has people to look up to. Meanwhile, a few men comment to one another that Eli has just had valve surgery. He is eighty-eight. One of them says he thinks it was a bad idea. He thinks it's much harder to go through than a bypass. Physical therapy does wonders, another few men decide. A forty-eight-year-old demonstrates a few exercises for an eighty-four-year-old who is annoyed by problems with his foot. Jacov doesn't look too well, some men mutter to one another. He had a bypass a few years ago, in his early seventies, and he's had a few balloon procedures since then. They're concerned that he's so pale.

Health is a constant backdrop for conversation and comment. The concern shown one another is a salve. When people inquire about one another's

Diamond Stories

health, discuss treatments, compare experiences, and offer reassurance and
ideas, they are all bathed in the value that each counts to the other.

Talk shifts to transactions with local traders, vital information for everyone who might someday be involved. Heads wave in distress—some more
surprised than others—about news of a trader who can't pay. Arguments
about Israel, New York events, the price of rough, and the current cultural
scene are all laced with lighter fare. Laughter spatters out of traders alert
for wry asides, sarcastic twists, and the darker underside. Interspersed are
the stories and jokes, many about age.

Lev recalls that years ago people used to claim to be different ages, depending on the situation. You couldn't trust the authorities, and you were
always trying to elude military service or whatever. Lev's father had three
passports. The others smile, recognizing this category of story. One showed
a birthdate of 1908, one said 1912, and the other said 1916. So if you
needed to avoid joining the army, you used one and not the others. That
recollection prompts Chaim to remember a joke. Pinchas always lied about
his age. Finally, someone asked him, "Pinchas, did you ever consider just
telling the truth and saying what your real age is?" Pinchas rubbed his
beard, looked thoughtful, then answered, "No, this I never considered."

Quiet laughter and deep nods greeted this story. Personal stories of es-
caping the Holocaust in the '30s and '40s or of using deception during the
pogroms in Russia earlier in the century were in everyone's immediate experience or family repertoire. The traders generally knew many of the stories about one another. A brief acknowledgment would signal recognition of
the painful past.

Though people complained about the other people in the business,
lamented the decrease in trust in the industry, and recounted the difficulties they had in paring the essence from irrelevant chatter or deception,
they seemed to derive genuine pleasure from the others year after year. The
depths of the friendships and the interlocking of the relationships were im-
pressive. One man in his late eighties still had four friends from kinder-
garten whom he would see a few times a week in the club or in the offices.
They occasionally marveled at this fact and took it for granted. Despite the
eruptive events of the past century that had sheared people from one another, they had continuity.

Entwined with the pleasure and annoyance of past ties was the game of
the work itself, sharp, wary, and ultimately serious. (Though having been
through so much, they knew in their bones that business is only money.)
Trading was a chance to spar, to display wit, and to appreciate the barbed
retort; it was also a display of competence against the accumulating years.

As one ages, the quality and circumstances of work change, and the
pace alters, they said. A tentativeness is typical as one starts out, making

contacts and learning the ropes. This period is often followed by a time of confidence and willingness to be bold—and perhaps to push the firm in unexpected directions. The pace in the older years is slower but steady. A person can be less involved in the business but not too detached. Finding the balance of being involved without being overwhelmed by the activity is key.

Allen, a trader nearing sixty, stressed that retirement had its appeals. Business is emotional and hard, he told me. Work is demanding. He said he hopes he can stay active in the business until he dies, as his father did. He hopes he feels up to it. As you get older, he explained, you get less willing to be bold. You want to hold back; you wonder: Why bother with all this buying and selling? The risk appears scarier to you. You don't look for such excitement anymore. When you have kids in the business, you keep going. They keep you going. Perhaps their firm would wind down and close up if there were no kids coming in.

Many people found that they were able to work, usually in modified form, into advanced old age. Though some men and women felt imprisoned by their fathers' continuing to order them around, others had worked out early antagonism and found pleasure working with their elderly fathers. As the children felt confirmed as adults and respected for their judgment, they also gained new appreciation for the knowledge and expertise of their fathers. Though some family firms split and failed, those that survived had learned to work and live with one another, perhaps like the diamonds that surmount ocean battering to emerge as strong gems.

◆

Like the Chagga of Kilimanjaro, the old diamond dealers live with their ancient enmities and enduring friendships, and they witness the entrance of vigorous new talent into their midst that challenges and flaunts its youth and bold energy. But being old, keeping relationships, and missing those who are gone lends a counterweight to other experience and finally is its own justification.

With the long and slow accretion of time, sometimes even the most bitter anger "wears off," as Leon related. Thirty years ago or so, he said, gesturing toward a man nearby, that man cheated him out of something like $20,000—not an insignificant sum. After being unable to get the money back, Leon refused to talk with him or to even acknowledge him. Ten or fifteen years later—he doesn't remember exactly—Leon returned a nod of the head when they were passing each other. Another ten or fifteen years later, they exchanged a greeting. By now they say hello; they don't have much to do with each other, of course, but in any case, the anger can "wear off" with time, as a stone gets worn down by water passing over it.

Of course, old hates also can stay inflamed, and old age does not necessarily make one mellow. Perspective can help.

With age and the perspective it may bring, change is necessary. How do you adapt the momentum of trade to your diminished energy with age? Some people are more successful than others. You need to stay involved, Allen said. You can't take major vacations, or you lose out on the scuttlebutt of what's going on, what the prices are, and what the feel of the market is. Perhaps the main factor is that you get *perceived* as not really active. You realize your heart is not beating real hard. It's like being a fish in the middle of a school of fish, and they're all swimming, and you're in the thick of it. Picture that versus sitting on the bank and watching the fish swim and observing that there's a lot of activity, but you're not quite in it yourself, he said. It's hard to be both in and out.

On the other hand, there was Ira, ninety-two, with no children in the business, who managed to stay both in and out, involved at his own pace. An exception? Allen wasn't sure.

Ira had gone to Belgium from Galicia in Poland in 1920. A year later he started cleaving. After hearing Churchill's famous speech in 1941 about fighting on the beaches, he declared, "Children, it's time to go." They went to Brussels, then Paris, Spain, and Portugal, and with Spanish visas and exit permits, they bribed their way to Santo Domingo—or they "would have ended up as ashes in Treblinka," Ira added simply. They came to New York in 1946, rebuffing Belgium's invitation to return, and he established an office. Your mind is a muscle, Ira told me. If you stay in bed, you get wobbly. If you don't use your mind, it atrophies.

However, not retiring has definite disadvantages, Ira allowed. He said he often wakes up and doesn't feel like going in. Then again, he can determine how often he goes in. Instead of going to the club some days, he calls a colleague to find out what's been happening. He said he doesn't have to be at the club to keep the sense of the marketplace. He can get the feel back pretty quickly when he returns. He is proud of his great memory.

His positive outlook on life was evident on his face and in his bearing. Sporting only the slimmest of elegant canes, Ira exuded the satisfied look of a proud man. Honesty was the most important factor, he said. You must be completely honest! To be less than totally honest is like being a little bit pregnant. You fool yourself if you pretend otherwise. He said he could look everyone in the eye and say, "I never cheated you." His mother told him that the fruit is only as good as the tree. All you have is your good name. But now rules are needed to safeguard honesty.

He told me a story: An old man planted a tree and was tending it carefully, nurturing the roots, hilling the dirt up just so around it. A man walking by asked him how long it would be before the tree gave fruit. The old man answered, "Sixty years." "So why do you work so hard when you won't get any benefit from it?" the other man asked. "Because others planted trees that I get fruit from that they didn't see, so I should do the same." This man then went to sleep, and when he woke up, he didn't know anyone because it was sixty years later.

His story conflated two stories and two morals, one about legacies for the future and the other about aging. That is the real curse of old age, he said, as he finished the tale: that you don't know anyone anymore. But the benefit of legacy was in the story, too: you create a path that others follow—and if you live long enough, you might see them do it. When he was done with this story, he was paged by the intercom. It was his wife, his "controller," telling him it was enough; it was time to come home. Off he went.

If you have inventory, you don't retire, one of the diamond traders told me. Ira related a saying: "If a diamond dealer dies with money to leave, that means he died before his time." You sell what you have, and you buy what looks good that you think you can sell. The fear of expulsion from the country you lived in, validated so many times in the past, demanded available inventory with which to flee. People stashed their diamonds, their concrete securities, in undetectable hiding places, ran, and continued a way to live.

"Aging is a moral and spiritual frontier because its unknowns, terrors, and mysteries cannot be successfully crossed without humility and self-knowledge . . . without acceptance of physical decline and mortality. . . . Individuals must be willing to persevere on the tragic journey to self-knowledge and communities must be willing to tolerate the unknown, the fearfully alien old person," Cole has written (1992:243–44). Heschel, a man around eighty-five, talked soberly about his diminishing circle: If I don't accomplish anything—that's the case more than before, you know—it has to do with the fact that my contemporaries are dwindling away, either by retiring or dying, he said. It diminishes my sense of accomplishment because they're not there anymore. The ones who are younger are the ones with whom I have no rapport. All those Hasidim—much, much younger. I'm trying to be honest. I feel frustration. It's a question of not being with my peers. My age group has died out or retired—in any case, they've disappeared. People like Arnold, he's three-quarters retired. And there are more and more. There aren't replacements among the younger ones. The younger ones are drawn toward their peers, who are people younger than my sons.

Business, like life, was combat, he declared. You get pushed aside, either by your own sons or by other people. People stay in as a struggle against that. They want to feel alive. You have to fight. If you can't do it anymore, then it's time to retire. We admit it. I see younger people come up all the time, and I can't do as well. Some people go every month to Antwerp; I can't do that. And the push wasn't the same when we were sixty-five and seventy as when we were forty-five. We were more powerful before. At sixty-five, I didn't want to kill myself. Other things were more important to me, the grandchildren. It's about relationships. It's a very varied picture, like a kaleidoscope.

The death of colleagues affects the traders. "When you get old," Bernard, in his eighties, said with disgust, "all your contacts—" he sputtered and flicked his fingers as if to shoo a fly—"Pih! Die!" One trader told me that he hadn't gone to the club since Shmiel had died; he missed their animated talks about politics.

Death was definitely not okay, but nothing could be done about it. Sometimes they reminisced about those not there anymore, but they didn't linger on the subject because they were rooted in Jewish and business practicality and knew that grief-tinged life was real and now.

◆

As diamond traders aged, they reconciled the demands of age with the pressures of work and devised various ways to modify their lives. In general, these older diamond traders appeared to have flexibility.

That the work of the diamond business can often accommodate individual alteration in routine lessens some of the sting of aging. Instead of being severed from friends, role, and workplace, as Savishinsky (2000) has described the mixed experiences of a group of American retirees, diamond traders are often able to retain desirable features of work. They age with their friends, the people with whom they have constructed meaningful lives for years and years.

One man in his mid-eighties continued to work about six hours a day. He arrived at 9 and left around 3 or 3:30. He did not need to sustain great momentum to know the prices and the marketplace and to be perceived as active. "This is something your nose remembers," he said. "You don't lose that."

A common strategy was to vary or to shorten the hours worked. Their hours ranged from full-time to only a few hours a day; in some cases, they went to work only a few days a week. Bernard, in his mid-eighties, didn't trade in Antwerp or Israel anymore, as he used to, but still went to the club every day to buy and sell stones. Weather, illness, family needs, and whim were frequent determining factors about going to the office or the club. For many older traders, however, internal rather than external forces seemed to determine how driven they were about going to work. Individual personality factors seemed more influential than the constraints of the business. I felt that I could not ask, but the necessity of a paycheck did not seem to be a deciding reason.

Many diamond traders pursued outside interests. They traveled, participated in sports, pursued hobbies, and took courses alongside their work. As they aged, some of them expanded the time they devoted to these leisure activities.

The Jewish calendar, chock-full of holidays and holy days, determined when the club was open and thereby set the clock on the industry, at least in New York. Older traders often took longer vacations at these times of year. Slack times in the business cycle became vacation times, too. Like many retirees in the Northeast, many older traders became "snowbirds" and spent months each winter in warm places like Florida. Josef, an eighty-nine-year-old manufacturer, went to Florida in the winter. I don't retire, he said, and I don't think of retiring because I love my work. I come to Florida in the winter. I play a little bit golf. I do a little business with some customers. I go to people when they need a stone; I get them a stone. When they need a jewelry piece, I get it for them. I know a lot of people.

In New York he kept his same hours in the office, 8 to 6:30. He ran the factory and bought and checked the rough. He gave the rough to the workers, told them what adjustments he wanted, sent it back, and so forth. He bought finished goods from overseas as well and flourished with this lifestyle of work and vacation. In New York, he served as an arbitrator.

He told me: My son is now supervising. I let him do it. It's the right thing to do. That's why my wife says I work for him. Like Josef, other men who were no longer in charge of their firms expressed satisfaction that they still had a viable role but not the primary worry for the business.

A manufacturer of industrial diamonds said that at age eighty-two he thought his ability to sort diamonds was unimpaired with age. Instead of working from 9 to 9 as he used to do, he sometimes felt fatigued earlier and therefore worked from 10 to 4. He was not as intensely interested in the worldwide political aspect of the diamond business as he used to be. As you know, he told me in 1995, the Argyle mines in Australia are resisting going with De Beers. It may be that there will be two major forces—De Beers and Argyle—to contend with, not just De Beers anymore. Ten years ago he might have been upset about this, but now he shrugged. He still has interest in the work, he was quick to clarify. If the interest wasn't there, he would quit. He is simply less worried about how the geopolitical situation will work out. He derives tremendous pleasure from the work. He enjoys the sorting. He's also proud to work with his daughter—the fifth generation.

Younger people in the club did not discount these men, perhaps because they could see their futures in the older traders and imagined their own diminished roles one day. Maybe they approved of the blurry line between those who were retired and those who were not, seeing the line as a good way to remain members of the DDC and to stay a part of things. Perhaps, too, they had respect for those old men in the back playing cards. Some of the cardplayers were finessing diamond deals and the stock market with skill and know-how like an old tennis pro who conserves his energy by placing his expert shots just so.

Chaim gave the example of the man who plays chess in the back of the club every day and asked what I would "call" him. This week, Chaim said, this man bought a stone valued at $500,000 in partnership with another man. Is this man *retired*? You could be a dealer and stay home much of the time, with brokers out showing the stone, negotiating, knowing the prices you'll accept. Your merchandise is in play, and therefore so are you. *You* decide whether to accept an offer and whether to agree to different terms. You can be flexible with your time and with your physical presence, and you can make this arrangement work for you.

◆

For many, working is a way of continuing a lifelong routine that constitutes a lifetime identity. None of the people I spoke to were elderly women, so I do not know how they felt about continuing to work into old age. The men I talked to regarded not working as a kind of not living. Manhood and life were part and parcel of work. The concepts fit together and composed each other. The meaning of work did not necessarily change over this life course. For these men, the continuity of work combined with the continuity of life, creating meaning (see Barth et al 1995, Rimer 1999).

A manufacturer in his mid-seventies, Stefan, told me that the meaning of work for him had remained constant. "I'm still working," he said. "The meaning hasn't changed. Right now and before, it's always been the way of making a living. There comes a time when you think you should retire, but you don't. Why? Because I wouldn't have other things to do. I used to think that I'd become a sex maniac, but. . . . Other people retire so they can do other things, but I do other things anyway. Men, when they retire, don't know what to do, and I wouldn't know what to do with myself. I get pleasure in my work when it's finished. For me, everything's a challenge. These are problems that need to be solved, and I solve them. When I solve the problem, I feel good. That hasn't changed. . . . The problems come to me on a platter, and I like to solve them."

Was he more patient nowadays? More patient, he answered (patiently). Why should I be more impatient? he then challenged me (impatiently). Did he work less now than he used to? Now he worked from 7:30 until 4:30, but he used to work until 7. He still took a bus or walked to work and refused to take a cab because, he said, "I'm a democratic man." He was not concerned about money, feeling he had enough to last him the rest of his life. "I want to keep working because I don't want to change my lifestyle," he said. He added that he didn't like to stay in bed in the mornings. He looked forward to each day's work. "I hope to stay working as long as possible," he concluded.

A number of men, older and financially secure, talked about the business as fun. Milton, a burly manufacturer in his early seventies, confident, almost swaggering, declared that he would not think of retiring because work was a game that was growing more enjoyable. "When you get older, your ambition isn't the same. You still have ambition, but there's less pressure and you don't take it as seriously." Milton had worked by himself for years. He said he would work as long as he could. He continued with his factory, working eight hours a day.

Marcel, a man around ninety years old, somberly told me, "I work because I don't want to get lazy. I'd be bored, I know. You know—like the couple that 'marries for better or for worse, but not for lunch'? That's what I mean." One concession he had made recently, however, was taking a taxi to the city from Queens rather than using the subway. His working less was painful to him, but necessary. Changes in the business were making it harder for him to keep up—the price list, the certificates, the electronic world impinging everywhere. With less work, he was less challenged. Plus, he felt he was witnessing a new world and didn't know whom to trust anymore. "Trust—they don't know anymore what your word is," he said sadly. He blamed the Israeli traders for some of the decrease in honesty, the decline of ethics. Where was the loyalty between the trader and the customer? "If it's ten cents less, you don't have a customer anymore," he summed up wearily, shaking his head.

Having one's children in the business had something to do with the choice to retire, two brothers in their seventies ventured. Their grandfather and father did little once the sons entered the business. Their grandfather did not know how to guide the business once the sons started to work. Their father became ill soon after they began to work, so health limited him. Both grandfather and father were titular heads of their firms.

The manufacturers believed that work was a way to occupy one's days, and they enjoyed the *schmooze*. Work might be worth continuing even if you took a loss, they said. They approved of a recently widowed neighbor in the next building who was around ninety; he went to work every day even though his son and nephew did the bulk of the work. Going to the office was a way to structure his life. As for themselves, they hoped for continued good health that would allow them to keep working.

Two men reading newspapers at the card tables at the back of the club were continuing a pleasant routine they had established long ago. One was telling the other some of the stock prices. In his late seventies with an ill wife, he worked part-time. "If you retire early," he said, "you get lazy. If your mind is working, why retire?" Then he challenged me suddenly, asking, "Well, how old is your uncle?" When I told him, he said, "Well, then," case seemingly closed. The other man commented that he liked to see the same people every day. "That's the most important thing!" he proclaimed. He had retired ten years earlier—"and I wasn't sixty-five when I retired," he added, flaunting his advanced years. He didn't trade anymore, but he got his mail at the club, where it was more secure. "At home they would sniffle around my mail."

When I asked a man pushing ninety if he and others talked about aging, he answered: We don't talk about it because it's self-evident. I know a lot of people who don't do anything—sit in the back and play cards, or come in for a half hour or an hour. Of course, it's primarily social for them. They have no children at home. Some of them are widowers, and then this is the sociability they have. Some of them have wives, but they're not so interested in carrying on conversations with them—and vice versa.

Josef, the eighty-nine-year-old manufacturer, believed working was the right solution for him, but he wouldn't make that same assumption for others. I don't know if most people want to keep working, he said. There are people in the club who are fifty-five years old who want to retire. I say it's a big mistake. . . . it gives them more time to argue with their wives. I say, let me work 'til I can't. I love it. I love that I'm busy and that my brain is working. I was working on a necklace for my friend, and she said: Josef, why don't you retire? I said: Because then I wouldn't be doing for you what I'm doing.

Many of the men expressed the common American belief that continuing to work was a way of staying healthy and mentally alert. The message that staying active equals successful aging and forestalls decline was probably the most unified idea I heard about work and retirement from them—almost as though they had read and agreed with this currently ruling gerontological mantra in the United States.

Josef continued: I was with my son and I told him a certain stone I remembered, and he asked how I could remember that. I remember so much. I do, thank God. You know why? Because I'm busy. Same thing with your uncle; he's that way, too. Same thing with my wife; she's busy. She remembers all the phone numbers.

Bernard, eightyish, was showing stones in the club to Menachem, a bearded Hasid in his thirties or forties. Menachem answered the question I lobbed to Bernard about work by saying: What's good about working? You can come in anytime, you can be flexible, it's great to work for yourself, and you can decide for yourself what to do. You don't have to take orders from anyone! Bernard waited his turn and then with great deliberation said: You Must Stay Occupied. Menachem pored over the array of small stones, loupe to eye, nose to parcel paper. It's all social, Bernard continued; that's the bottom line. The men in the back play the stock market, they play cards, they stay "in" in various ways, but mainly it's social. Even this younger guy, Menachem—he waved his finger at him vigorously—he doesn't work every day! Menachem listened as he pricked a diamond with the fatty part of his forefinger and placed it daintily on his hand, then inspected it anew with his loupe before wrapping it back up briskly in the crackling parcel paper and smoothly opening another.

Then Bernard posed the essential Jewish question, partly, I suspect to taunt the devout Menachem. "What makes a Jew?" he asked the ceiling, a question most Jews eventually spend a lot of time thinking about. Bernard said he didn't share the beliefs that many of the men and women here have, and he stopped to wave his arm across the room, glancing at Menachem, who still looked intently through the loupe. He himself never went to *shul* on *Shabbes* or on the High Holy Days, Bernard added, and he didn't agree with many of them politically or socially—but he was a Jew. Why? Because Gentiles would still beat him up for being Jewish, that was why.

Without comment, Menachem stopped only to wrap up the next stone in its parcel paper. He pushed it back to Bernard and opened a new one, inspected, refolded. Done, he pushed back his chair and stood up with an almost imperceptible nod while another Hasid with spindly white beard coiling down to his chest angled immediately to replace him in the same chair. Bernard tried to introduce us but this man waved the formality away as unnecessary or unwanted. He muttered something about knowing who I was. I asked recklessly: Why continue to work? The Hasid said, "You should ask Bernard." Bernard said he was not supposed to look at girls, but he did. The Hasid was writing out a check and handed it to Bernard. Bernard lit up. "Now you see why I keep working?" he said, pointing to the check. "This is why I keep working!" he crowed happily.

The Hasid added: And it's not physical work. No one here wants to think about retiring. They all want to work as long as they're not sick, and that's what they hope for. But also, Bernard insisted, the social part is great; you see people you've known for years. Bernard folded the check and put it into his pocket.

The Hasid left and I saw Heschel, who told me he had just sold a stone. He said: I sit here three or four hours a day in the club and the sale comes along just like that—a fluke, and it was done in a few minutes! Someone had asked him whether he had a certain kind of stone, he thought about it for a minute, thought he might have it in the vault, went downstairs and got it, and the buyer wanted it. Heschel had had the stone for at least seven years. His remembering the stone echoed Josef's belief that continued activity kept memory sharp and keen. I was witnessing it again: traders asking one another for a particular kind of stone, each of them rummaging through his mental catalogue to suggest various stones in his inventory that might be suitable. The traders made a show of this mental feat. The public performance of producing accurate information about specific gems left everyone with great satisfaction and renewed commitment to this ideology of thought in action. Quick thinking makes the man and maintains life.

Heschel talked about the joys he felt in the business. I still enjoy it, he said—feeling adequate, making the deals, either buying or selling. I have to know why I'm getting up in the morning, he said. I think: What's going to happen today? Who am I going to see? What calls am I going to make? Every day is new with possibilities. On Friday, I saw a big stone, but I ended up not buying it because I wanted to go in with a partner. . . . I showed it to my partners. One was interested, the other lukewarm, didn't think it had enough color. We showed it to someone who might be with us in partnership. I thought we could make a profit on it, manufacturing it. So every day it's something different. The other day something similar happened. It's very exciting that something can happen every day. That's been my life for many years. It's challenging. Other people feel very much the same way.

Many of the men and women, in fact, told me they felt very much the same way.

[8]

The Courtly System of Arbitration

Settlement by arbitration is a meritorious act, for it is written:
"Execute the judgment of truth and peace in your gates."
(Zechariah VIII:16). Surely where there is strict justice there is
no peace, and where there is peace there is no strict justice. But
what is that kind of justice with which peace abides? We must
say: Arbitration. So it was in the case of David, as we read: "And
David executed justice and righteousness towards all his
people." (Samuel VIII:15). Surely where there is strict justice
there is no charity and where there is charity there is no justice.
But what is the kind of justice with which abides charity? We
must say: Arbitration.

(Talmudic passages quoted
in Schumach 1981:51).

Alongside handshakes and the uttering of *mazal* to finalize a deal, the arbitration system of the diamond industry is the crowning glory of the trade. Arbitration originates from Jewish law dating at least from the first millennium. Used in other "Jewish-oriented" businesses, like the garment and fur industries, in the United States, arbitration is one of the features of the diamond business that illustrate the ancient roots of the industry as well as its adaptation to modern times and globalization. Despite encountering increasing strain, arbitration works extremely well.

Talmudic prohibitions against going to non-Jewish courts, Jewish writings stressing the internal resolution of disputes, and the *Shulhan Arukh* (the Code of Jewish Law) stipulated that secular courts could be used only if the participants had decided not to abide by the ruling of the rabbinical court.[1] When non-Jews (Gentiles) were unwilling to go before a Jewish court, however, Maimonides allowed in the twelfth century that the Jew could go before the Gentile court.

Jews have participated in arbitration in their own communities throughout the world. For example, Shirazi Jews in Iran settled disputes by arbitration to avoid the loss of honor that would result from taking a fellow Jew to court. A guild elder usually mediated in a business dispute as Loeb described in the following passage: "When the mediator comes to what he considers a fair decision, he pronounces it to the disputants and the angry words die away. He holds the head of each disputant with both hands and kisses him on the forehead. The parties then kiss each other on both cheeks and the dispute is over. More serious disputes require several mediators" (1977).

The *Beth Din* in Antwerp and elsewhere handled disputes using Jewish laws attuned to the particular disputes at hand. Not only were the clashes hidden from the prying eyes of the surrounding community; many Jewish conflicts required Jewish understanding. A memoir provides this glimpse into the pre-World War II Greek experience: "All is very quiet. Grandfather is discussing with other members of the *Beth Din* a forthcoming trial. Few Jews use a secular tribunal. They come to the *Beth Din* and they obey its decision. This is a matter of survival in the Jewish community, for it is the practice among the merchants to borrow money or enter into agreements with one another on the basis of one's word or a handshake. No documents are used. Because the Greek language is foreign to most Jews, they avoid the 'regular' courts altogether, but apart from this they prefer to keep their problems 'in the family'" (Molho 1994:2).

The same is true in the United States. "The reluctance of Jews to air internal disputes in public was, throughout, a prime motivation in the New York Jewish community," Goldstein has written. "Many people presented their cases to us [the Jewish Conciliation Board] instead of to the civil courts, in order to avoid *Hillul Ha-Shem*, the desecration of God's name, and to safeguard the good name of the Jewish people" (1981:98). Insecurity about their position within the larger society often led Jews to avoid the scrutiny of outsiders and to "find ways of settling disputes among them-

---

[1] The following sources were helpful on Jewish dispute resolution in various locales and times: Mintz 1992, Goldstein 1981, Loeb 1977, Gutwirth 1968, Howe 1976, Weisser 1989, and Mitchell 1978. I am indebted to William O. Beeman for directing me to the Loeb source.

selves short of open battle; [they] had moral obligations to one another that would at least soften the ravages of class conflict" (Howe 1976:302).

The organization of the *kehillah* in New York helped immigrants both adapt and continue Jewish life. It "set itself up as a self-supporting clearing-house for Jewish welfare, as well as a representative of Jewish interests in the society at large. In other words, it was expected to function both as the arbiter of problems internal to the structure of New York's Jewish community and as the watchdog over Jewish concerns vis-à-vis the city" (Weisser 1989:145).

The marginality of Jews was central to the evolution of this strategy in the United States. It was "very important to look good in the eyes of the Gentiles" (Morawska 1996:196). Cousins' clubs, family circles, and *landsmanshaftn* were Jewish self-help groups that aided life in New York in particular. Often conflict-ridden, these groups resolved disputes and sometimes charged fines.

Today's New York Hasidic community uses the secular court only when the religious court is unable to enforce its decisions. Not doing so brings notoriety to the group and constitutes a snub to the authority of the *rebbe*.

Arbitration in the diamond business likewise is a time-honored mechanism for group cohesiveness as it smoothes schisms and conflicts. It is an emblem of group distinctiveness as it maintains separation from American society. As Ilani et al. have raved, "Behind the business deals made by telephone, by telex or by facsimile, stands the well-seasoned wisdom of an ancient people [that] accompan[ies] the Jewish businessman in the diamond industry in Israel and throughout the world, in the past and as in the present (1991:20).

However, the insistent penetration of the U.S. legal system offers modern challenge to its effectiveness, and as much as arbitration resolves conflicts, it also reflects division.

An arbitrator for many years offered his summary of the changes of the last thirty years:

We allowed arbitrators complete leeway in the past, he maintained. They could conciliate, and they didn't let lawyers in. One day a member said he wanted his lawyer in, and the administration realized they couldn't deny his request. So lawyers were allowed in. Then lawyers started badgering arbitrators during the proceedings, so then the arbitrators wanted the club's council to be present during the arbitrations. Then they wanted the club president to be a lawyer. Then they decided that the arbitrators should be educated, so now they always have some kind of course about arbitration, the rules, the laws, and so forth. Then they realized that there's a whole body of law out there in the state and the nation that governs arbitration that they should know about. They realized that they had to conform or their rulings could be overturned in court for either misconduct or procedural errors. Otherwise, they have incredible leeway still. For instance, they can decide what evidence they'll hear. Finally, he concluded his summary with his assessment of arbitration's present status. These new procedural rules may rankle the old-time people, and compared with the rules in other bourses, these rules are more constraining. Still, arbitrators have lots of discretion. In short, it's a system that works really well.

What arbitration is, and how and why it works despite its new pressures, are described in the following pages. That governments and industries are looking to arbitration in the diamond industry as a model for resolving disputes outside courtrooms—in domestic law, labor conflict, traditional civil suits or within and between industries—is testament to the resiliency of this institution.

◆

A writer for a diamond trade magazine told me that much of diamond trade conflict is personal. It's like the Ed Koch show on TV, he said, with one side telling his story and then the other side having his turn. Often the problem is no one's fault—something got lost in the mail, that kind of thing. Is there agreement or not: is it *mazal* or not a *mazal*? "Traffic court stuff," he concluded. As a former DDC official put it, "Trading diamonds can be as simple as 'mazal and brache,' an exchange of diamonds and an exchange of money. It can also become mired in complications, misunderstandings, unforeseen problems, or outright bad faith" (Lubin 1982:20).

In all contract disputes without written contracts, opportunities for dispute are ubiquitous. The terms of a deal are disputed. A trader changes her mind about the deal and attempts to back out. Payment wasn't made. Conflict involves whether payment was made or whether a stone was returned. The wrong stone was returned. The terms of a partnership are disputed. The deal is off, but the broker claims he should be paid. A partner did not properly protect the stone while under his care. Was a partnership in effect when the deal was made? Was the partner properly consulted when the deal was made? Whose responsibility is it that the stone was lost or stolen? How can creditors be made whole after a bankruptcy? What responsibility does the broker bear for a lost or stolen diamond? While the complications are endless, the cases boil down to several key principles, as a seasoned arbitrator summarized with elegant simplicity:

The issues don't change. It's always A sold to B, and A says that B didn't pay him. Or, you took the stone, and you didn't sign the memo because you were in a rush and said you'd remember, and you know that's true because we all do that. We all know what that's like: we intend to remember. So, the issues are always the same, and they always have been. And human nature is always the same. We look at the litigants and try to decide who is telling the truth, and we think one of them is. We've heard the stories a thousand times, but figuring out the puzzle each time is different, listening to the tone of the story.

A legal analyst observed that while cases often have either explicit rules or custom-honored remedies, complex cases defy prediction as the arbitrators attempt solutions and redress (Bernstein 1992). An arbitrator explained, "Every case is a variation; every case is different—different people, different amounts—but it's always the same thing: it's one word against the other, very often with no evidence, and we have to rule by hearing the arguments on both sides and judging with our gut feeling—who's lying, who do we think is lying, or who is lying more, or, you know. It's very difficult. . . . We judge a person's character, we investigate the evidence, and we hear the story. We have already decided many cases against people that we like, for instance, and found for people that we don't like. This is the essence of justice. . . . I've made plenty of enemies in time."

Farber described a sample case in which two stones of the same shape and weight but of different quality might be switched, with one sent to a customer. If this simple mix-up were not immediately remedied, an argument over who should take the loss could be arbitrated (1996:63–64). An arbitrator described a recent case to me.

"[The case] was a firm represented here that sold to a particular firm a certain amount of goods. . . . [The] supplier was an Indian firm wh[ich] dealt in cheap [inexpensive] goods, and the buyer was an American firm. At the end of this phase of dealing with him, the Indian firm sent in, I think, 10,000 carats of cheap goods, and the reps in New York gave it in consignment to the American firm. After about three to four weeks, the Indian firm went ahead and billed it to the American firm. The American firm returned the goods, didn't want to buy it. The Indian firm maintained that they had sold the goods. So this was the litigation." He paused, then went on. "American firm maintained that they only took it on consignment, and they examined the goods, and they didn't want it. The Indian firm said no, they always had that kind of deal with them, that they gave it to them on consignment for a day or two and then billed the goods." He looked at me to see if I was following the sequence. He then summed up the problem. "And there we have different kinds of evidence to examine: How long did they keep the goods? Were the goods the same as the usual type which they had? We asked the rep, who didn't want to commit himself. We had to judge by gut feel. We didn't have any evidence for or against. They had dealt with each other for the last two years. We examined that history, we examined how they had dealt with each other, [and] the time they had left it in consignment with this firm before they billed it, before they told them to make the bill. And we came to the conclusion that they shouldn't be forced into buying the goods. [It was fitting the pattern.] They accepted the decision. They didn't go on appeal."

◆

Such cases are handled by a standard arbitration process. The emphasis is on conciliation and resolving the disputes internally and rapidly. DDC membership requires submission to arbitration instead of to outside courts.[2] According to Bernstein, approximately 150 cases are brought to arbitration in the DDC each year, of which about 85 percent are resolved in the prearbitration procedure mandated by the club (1992:124).

Incentives, such as giving refunds on arbitration fees if conciliation occurs prior to the formal arbitration, are offered throughout the preliminary process so that most conflicts are resolved early on. In one case, someone approached an arbitrator and asked whether his situation qualified for arbitration. A partner had promised him that a certain pink stone would turn a profit. When the GIA certified it as purplish-brown rather than pink, the other person backed out of the deal. After listening to the story, the arbitrator advised the men to resolve it between themselves rather than proceed to arbitration.

In another case that didn't make it to arbitration, a man had a stone from a dealer in Mexico. He gave it to someone, but it was lost and subsequently exchanged. It had been worth $2900, but the other person was claiming it was now worth $2200. The arbitrator urged them to compromise on a figure around $2500.

Arbitration is performed in accordance with Section 7501 of the Civil Practice Law and Rules (CPLR) of the State of New York (DDC By-Laws, Art. XII). Arbitration can also be used to settle conflicts between members of diamond bourses affiliated with the WFDB; sometimes arbitration can also be used when a member has a dispute with a nonmember. DDC members are required to abide by the arbitration decision.

There are two conflict resolution processes, involving a floor committee and a board of arbitrators. Forty arbitrators, including sixteen chairmen, are chosen for a two-year term on the board. Twenty-six of these are elected by

---

[2] While earlier DDC bylaws stipulated that members must submit to adjudication within the club, the 1999 revised bylaws added the following: "No member shall prosecute or commence any suit, action or proceeding against any other member touching upon any of the matters covered by Arbitration pursuant to these By-Laws or seek any remedy in Court except to confirm an Arbitration award. A violation of this section by a member shall result in disciplinary action against the member including suspension and/or fine" (DDC By-Laws, Article XII, Section 1a, 1999).

the DDC members; fourteen are appointed by the board of directors. Arbitrators are generally held in extremely high esteem. As a DDC official explained, "Their qualifications are experience, common sense, a good understanding of the business practices in the diamond trade and being well-versed in DDC bylaws. They must also have impeccable reputations" (Farber 1996:64). One arbitrator told me why he had been an arbitrator for so long.

"I've been doing arbitration since '57. It's a small service I've given to the club. I do two to five cases a year. Some cases run for some time, but usually a case is finished off in one or two sessions, and a session is one or two hours. [I got into it] because I was interested, and I was nominated. I didn't know I'd be good at it. I was interested in the problems, the solving of the problems. I'm still interested in those. It's as interesting as reading a mystery novel and finding out who did it. That's a crude example. But finding out what happened and finding out the ethics in question [is interesting]."

All new arbitrators attend three one-hour seminars at the DDC on arbitration procedures to conform to DDC bylaws, the CPLR of New York, and standard arbitration judgments. To initiate arbitration, one member writes a complaint to the managing director of the DDC. If the vice president of the DDC determines that the case should proceed to the floor committee, the floor committee determines whether a material issue of fact exists for arbitration to decide. The committee has the authority to exclude a member from the trading hall for up to twenty days and/or to impose a fine of up to $1000, decisions that can be appealed. When a case is not accepted for review, the parties are free to go outside the club for remedies.[3]

Each arbitration panel of three members, including a chairman, and appeal arbitration panel of five members, including three chairmen, used to be chosen by lot, but the DDC vice president now selects them. The composition of each panel is made public to the members. The vice president can excuse members from the panels for cause, or they can recuse themselves if they feel they cannot "fairly and faithfully act as an arbitrator in a specific case."

When arbitration occurs, the panel listens to the case and decides it

---

[3] In addressing a dispute arising out of a representation agreement between diamond dealers with respect to triangular shaped brilliant-cut diamonds, the DDC declined to arbitrate aspects of the dispute involving allegations of trademark infringement and unfair competition. Those aspects of the dispute were decided by the U.S. District Court for the Southern District of New York. In its decision, the court spoke approving of the DDC's handling of the matter: "The Diamond Dealers Club wisely refused to involve itself in the dispute concerning the trademark registration and alleged infringement." *Finker, Inc.* v. *Schlussel,* 469 F.Supp. 674, 678 (1979), *affirmed* 614 F.2d 1288 (2d Cir. 1979).

within ten days of the hearing. If the arbitrators are unable to decide by majority vote, a new panel is called. During the proceedings, the litigants present their sides of the case, offer evidence, and cross-examine witnesses. An attorney may represent litigants. Arbitrators are empowered to investigate the complaints, and witnesses are often called. The arbitrators may impose fines when litigants fail to appear, fail to produce documents pertinent to the case, or exhibit "contemptuous behavior during the process of a hearing." One arbitrator described the arbitration process to me as follows:

"Very simple. . . . The chairman leads the proceedings. He usually asks the questions unless one of the arbitrators wants to ask. Usually he directs himself to the chair, and he can ask the questions also. But it goes to the chair. There's a format like in court, I suppose. The plaintiff starts by pleading his cause. He gives the complaints and then there's the response by the other side. And very often there are lawyers with it. And they sometimes take over and explain some aspect of it. But it's the chairman who leads the proceedings. Each gives his argument. We ask questions in order to clarify certain pieces of evidence or lack of evidence. And there's a cross-examination. Each party can cross-examine the other litigant. At a certain point, when we think we've heard all the arguments, we stop it. The chairman decides everything. The chairman has a lot of power. He can decide.

"Then we usually dismiss the parties, or we let them wait outside while the arbitration panel deliberates. We discuss it among us, and we decide whether we want to hear them again, whether we have enough evidence, or, you see very often, we ask them to bring other papers that they haven't brought yet, so we have to have another session. And very often there are witnesses that have to be called that can only be called next time, so we have to postpone it again—people who can testify to certain aspects of the case, to the facts. We have certain cases that run through a year, and we have cases that have ten to twelve sessions because the evidence was lacking, and we had to come back and come back 'til we finally terminated it at a certain point. And when we've got enough information, we discuss it among us. If it's an appeal case, it's five of us. So we deliberate and we come to a solution, and it's majority rule. We don't decide on precedent."

A small part of the process is public. Names of litigants in upcoming arbitrations are posted in the club and the judgments are announced there, but the reasoning behind decisions is not included, and the arbitration proceedings themselves are secret. No notes are made, and no audio or video recordings are permitted. All the requests I made to sit in on arbitration hearings were summarily refused. If a judgment has not been complied with and an appeal has not been filed after eleven working days, the award and a picture of the noncompliant member is posted on the bulletin board of the DDC.

Though decisions do not rely on precedent and are not written down, individual arbitrators tend to remember their decisions even though the members of the arbitration board change. In this way a considerable legacy of oral history builds up, and members share and draw upon it. Part of this legacy is lost when a member dies.

Though the judgment of the arbitration panel can be appealed, the fees to do so are steep.[4] The appeal arbitration panel acts with the same powers as the arbitration panel. An arbitrator described it to me as follows:

"With appeal it starts de novo—completely different from the court system. An appeal with the courts is only an examination of the judgment by an appeal court, whereas an appeal with us means a new case, new arguments, new evidence, new people, new arbitrators. Very often someone goes on appeal, and he gets fined $10,000 in the arbitration. He goes on appeal [and] sometimes the appeal arbitration board will fine him $15,000! It may happen. They examine all the questions again. It's a risk for him. He has to make out a check for the first arbitration panel before he's accepted for appeal. There's no other level of appeal. This is it. [But t]here has been exception to that where somebody brings evidence that an appeal member was bribed or something like that—this is an exception. He can go to the board of directors; he can ask for a hearing with the board, bring his arguments with evidence. This has happened. They'll look again."

Arbitrators do not have to justify themselves, and decisions are unpredictable. They are based on common sense, trade custom, principles of Jewish law, and other legal notions widely accepted by the community. Arbitrators calculate damages by assessing the stone and the circumstances in accordance with those principles. Arbitration is not used frivolously; it is reserved for the most serious matters. The arbitrator continued:

"It's not legalistic, but commonsensical. Penalties and restitution is included. This is the main issue of the thing. We decide the amounts to be paid, not paid, the terms, et cetera, et cetera. . . . We are not called in order to judge the behavior of a person, just to straighten out differences between two people. This is the role [of] the arbitrators—very practical. We give our decision in terms of money or terms of no payment, whatever, but we don't enforce it. Our job is finished."

The members of a family firm described an example of how an appeal

---

[4] "A written complaint of the aggrieved member's claim shall be signed and filed with the Managing Director together with supporting documentary proof of the claim, if any, and the arbitration fee in the sum of one hundred dollars for the first thousand dollars of any claim or fraction thereof and ten dollars for each succeeding thousand dollars or fraction thereof up to ten thousand dollars and five dollars for each succeeding thousand dollars which filing fee shall be paid at that time. The claimant shall also sign a demand for arbitration and such other papers requested by the Executive Office to process the claims. The Panel of Arbitrators as part of their award may charge all or any part of the Arbitration filing fee to any litigant but may not refund such fee" (DDC By-Laws, Article XII, Section 2, 1999).

case reinstates the arbitration process from scratch. They had been defendants in a case during the past year and lost. Then the plaintiffs appealed because they felt they didn't receive a large enough award from the arbitration. The plaintiffs won again, and this time the defendants had to pay even more. Though the defendants felt they were in the right, they shrugged as they said so. That's the way it goes. It is part of the system to which they have subscribed.

Punitive damages or fines are often assessed against the person who has been judged to have behaved unethically or unprofessionally. Here, too, the type of penalty is unpredictable and varies greatly. A person can be required to pay a certain amount to a charity. One successful female litigant, for example, told me she wanted a charity donation and a personal apology because she had been insulted as a woman and as a professional. Bernstein provides another example:

"In one case, a dealer falsely accused another dealer of stealing a stone. The accuser subsequently remembered where he had put the stone and apologized to the other dealer. As the incident had become widely known throughout the club, however, the wrongly accused dealer brought an arbitration action against the owner of the stone for impugning his good name. The board ordered the man to make a full public apology and a $50,000 donation to a Jewish charity" (1992:127).

Schumach reported that when a member screamed at another member, "Hitler didn't kill enough Jews"—supposedly because the other member got a better seat at a table for viewing diamonds—he was fined $25,000 for his insult (1981:41). This case was remarkable in that it did not even involve a sale of diamonds.

Losses suffered in the diamond business that result in arbitration are usually never fully restored, yet any restoration seems better than nothing. An arbitrator described an appeal of a robbery case in which three large stones were in escrow; arbitration concerned how to divide the losses. In another case, the dealer had given out some goods that were then lost. The insurance company refused to pay for the "mysterious disappearance." In the arbitration the diamond dealers each took a large loss. The dealer was philosophical, commenting that these things happen.

Bankruptcies are complicated, and when many people are involved, there is pressure to restore some of the losses of those affected. When a person "goes sour," "gets into trouble," or "gets busted," the ripples fan out into the far corners of the diamond industry. People along the chain count on payment to meet their own obligations. I witnessed the "buzz" when news circulated that a respected member was in trouble. Akhil had given the man stones that he hadn't gotten back yet. He worried that he would never see his stones again. Others were similarly concerned. During the following days, people compared notes. In cases like these, multiple arbitrations disentangle the debts and minimize the damage. Creditor committees are formed. It'll get nasty, predicted a seasoned arbitrator about this developing case, and it'll be hard to sort out. Because everyone knows someone who is affected, confidence in the trading structure suffers.

When a bankruptcy is considered hopeless, little can be done. In a procedure known as the "knockout" system, the firm's diamonds are strewn on a table for the partners to bid on (Schumach 1981:38). Dealers traditionally had to agree to pay a certain percentage of their debt to their creditors until they had been paid in full, but some recent bankruptcy arbitrations have allowed greater leniency.

The arbitration process is considered formidable by most of the club members. A diamond dealer lawyer was attempting to calm her client prior to an arbitration. I heard her counsel on the phone: "Ezra, if you're that nervous about the arbitration, maybe we should settle and finish it." She told him he had to show up, that being a "no-show" would look extremely bad. She repeated that perhaps he should settle the dispute instead. He showed up for the arbitration.

Arbitration is solemn and feared by many. A dealer told me that his eighty-nine-year-old father-in-law, Yossi, was involved in an arbitration. Someone had his stone that was substituted for another when it was time to return it to its owner. Yossi had to show up as a witness. He had just had a pacemaker installed. The doctor monitoring the new pacemaker later asked, "What was going on between 1 and 3 P.M. that day?"—clear evidence that Yossi had been tense.

The story was important for another reason: the person intended to escape blame by claiming that he never had dealt with Yossi. But Yossi brought a memo saved from years before that proved the two of them had done business with each other. The first man's loss of credibility allowed the other man to win. It may not have been a legally binding document, but the memo was vital to the arbitration.

Arbitration is serious.

◆

Arbitration is still the only way to go, a diamond dealer told me. It's a great system. Of course, there are varied views about it. The diamond business is a $7 billion global business and it's much more diverse than ever. The legal system could not possibly handle it. It's too slow; it isn't knowledgeable about the business. Arbitration still has great clout among its members. It works.

Why is arbitration preferable to an outside civil court? Arbitration protects the group, those in the industry say. Creoles in Sierra Leone and businessmen in London have kept their affairs private to ensure survival and to reduce vulnerability, as Cohen has shown (1974). Group protectiveness is effective for weak as well as privileged groups. Secretiveness enables a group to conserve what it has—whether it's abundant and prestigious or scant and meager—against the onslaught of real and imagined threats from the outside. Arbitration in the diamond business protects both the vulnerable and the powerful.

Beyond these reasons, nonlegal sanctions in general help participants maintain reputation. Though business commitments are often unspecified or informal, Charny writes, "nonlegal sanctions—the desire to maintain reputation or profitable relationships, the concern for standing among peers, and the force of conscience—induce the promisor to keep his commitments" (1990:375). Shame or the fear of shame is important. Public posting of a noncompliant trader's picture is severely damaging and has a deterrent effect. Were written contracts widely used in the business, traders would still need to research the reputations of their potential trading partners.

Certainly arbitration is much more efficient than the courts, and civil courts sometimes harm rather than benefit litigants. When breaches occur in the trade, dealers and brokers can recoup losses and redress wrongs with less delay and more effectiveness via arbitration. Money is not tied up for a long time awaiting judgment, a crucial consideration in a business dominated by credit transactions. Unfamiliar with the diamond business, the courts do not fundamentally understand it and cannot gauge the nuances of trade norms from conflicting witnesses. Because of their lack of in-depth knowledge of the diamond business, the courts have difficulty determining equitable compensation for lost profit and calculating losses accurately.

The value of speed is important enough that arbitrations often proceed even when a litigant fails to appear. Also, arbitration avoids the high litigation costs of the formal legal system. Despite the benefits of arbitration, however, a person who would bring to light an opponent's breach of contract would himself suffer a loss of reputation because of going against the norm of privacy in business dealings. One dealer told me that going to arbitration represented a breakdown in the communal spirit. People figure they should be able to handle disputes among themselves without resorting to arbitration. It takes a lot to go to arbitration.

Arbitrators understand the basics and the nuances of the trade, and judgments are therefore appropriately consistent with the trade's values. Customs and frequently used trading routines are well understood. Arbitrators can use their knowledge of traders when they evaluate the litigators, judge the stories, and determine the outcomes. In addition, the value placed on conciliation encourages traders to settle their differences throughout the process. This value aids the smooth flow of business since it promotes the restoration of normal business.

The privacy of arbitration preserves dignity and minimizes damage to reputation if compliance is met. When the names of litigants are posted, diamond traders are protected by being made aware of who is involved in arbitration. Such information allows traders to discern appropriate prospective partners and customers. Since the extreme measures of suspension and expulsion are rare outcomes, arbitration preserves the viability of the community of traders.

Though lost profit is difficult for arbitrators to judge, they have more knowledge than courts about the kinds of losses that are likely or typical. Furthermore, they try to compensate the person who suffered the loss in order to redress the wrong.

That the arbitration panel can make a judgment in a brief time alone compensates the victim far better than if he or she had sought relief in the courts, which can take years. Most diamond dealers do not have ready access to credit or cash to make up for lost profits during the period of the unresolved dispute. They can go out of business waiting for a judgment, given the tight cash-flow margins under which they operate. Furthermore, were the dispute to be addressed in the courts, other dealers would view the litigant as a higher risk and might avoid dealing with him or her or might charge him or her steeper credit terms.

The reliance on reputation and arbitration continues as the industry expands, diversity increases, and new bourses are created worldwide. The use of technology helps transmit information about reputation more efficiently and reliably to all the bourses than previously, and arbitration continues to be effective.

◆

For the most part, New York courts have upheld arbitration decisions, and arbitration is increasingly touted as a model for mediating disputes outside the business. Not wishing to further burden the overloaded court system, the New York courts have minimally interfered in diamond business matters and have upheld the process as virtually sacrosanct. A dealer told me arbitration is a potent tool. It's rapid and effective, and members must agree to submit to its authority. New York favors arbitration and has incorporated principles from its system.

A decision in a DDC appeal arbitration may be appealed to the New York State court under New York law, but overturned decisions are rare and are based only on procedural irregularities. The New York courts uphold the validity of arbitration rulings because the standard of review in arbitration cases is narrow. Generally, awards may be vacated if denial of procedural due process, fraud, misconduct, or partiality or misuse of power by the arbitrator is found in the arbitration review but will not be vacated for mere errors of fact or law by the arbitrators. See New York Civil Practice Law and Rules section 7511. The New York Court of Appeals has held that the arbitrator "may do justice as he sees it, applying his own sense of law and equity to the facts as he finds them to be and making an award reflecting the spirit rather than the letter of the agreement. . . ." Matter of Silverman [Benmor Coats, Inc.], 61 N.Y. 2d 299, 308, 461 N.E. 2d 1261, 473 N.Y.S. 2d 774 (1984).

New York judges have encouraged litigants in commercial disputes to seek resolution among their peers. As long as arbitrators act within their jurisdiction, there is no need to interfere. In another case, the judge issued the following:

"Courts are reluctant to disturb the decisions of arbitrators lest the value of this method of resolving controversies be undermined. . . . Precisely because arbitration awards are subject to such judicial deference, it is imperative that the integrity of the process, as opposed to the correctness of the individual decision, be zealously safeguarded. . . . Arbitration by its nature contemplates a less formal environment than the judicial forum . . . and accordingly, arbitrators are not held to the standards prescribed for members of the judiciary . . . but must take a formal oath . . . and ought to conduct themselves in such a manner as to safeguard the integrity of the arbitration process. . . ." (statements from Goldfinger v. Lisker, 68 N.Y. 2d 225, 500 N.E.2d 857, 508 N.Y. 2d 159).

Instead of being concerned with content per se, the courts are concerned with procedure. A judge would be concerned that everyone understand the proceedings rather than with the content of what was said, for example.

A decision that was reversed in court was the Goldfinger v. Lisker case in 1986. Arbitrators awarded one member $162,976, after which the decision was confirmed in the courts. When the decision was appealed, the court reversed the original decision because a private communication between an arbitrator and one of the litigants related to the other litigant's credibility. That the other litigant did not know of or consent to this communication constituted misconduct sufficient to vacate the arbitration award (Goldfinger v. Lisker 68 N.Y. 2d 225).

A longtime arbitrator told how he learned to conform to the legal emphasis on procedure. Arbitrators should not allow anyone to leave to take a business call, for example, unless the proceedings stop. If proceedings continued, he explained, it could later be claimed in court that not all the evidence was listened to or that the litigant didn't have a chance to explain everything. New evidence or testimony could then be the subject of the court case. So at the end of his arbitration, the arbitrator says something like, "Is there anything that anyone would like to say? Do we need to meet again?"

But enforcement of arbitration decisions is more problematic.

◆

Enforcement is not easy. As one arbitrator said, "Could he be suspended or kicked out if he does not live up to the decision? We actually haven't had that. . . . It wouldn't be up to the arbitration panel to reach a decision of suspension. That would be up to the board of directors. Would arbitrators recommend suspension? I don't remember any such case. Enforcement is the job of the directors."

Though DDC members are enjoined from seeking redress in outside courts in most cases, the ability of the club to enforce arbitration decisions has been tested in recent decades. In fact, enforcement of *Beth Din* decisions may have been problematic in the past too, as this excerpt suggests. In Antwerp "the [diamond] Exchange members prefer to make use of the national courts of justice only in cases which call for the exercise of coercion (police, prison, etc.), and even then they do so generally after obtaining the consent of the Exchange executive," Gutwirth wrote (1968:133).

Notwithstanding difficulties in enforcement, the specter of arbitration has a bracing effect on the behavior of club members. When an unsuccessful litigant fails to comply with an arbitration decision and this information is posted on WFDB club bulletin boards worldwide, the damage to the person's reputation is serious and immediate. In addition, members know that expulsion or suspension from the club—though rare and sometimes reversible—is a threat the board of directors can use. Further, members understand that the Jewish rabbinical courts may also excommunicate a member from participation in Jewish communal affairs in the most extreme cases.

However, members also know that the club is increasingly loath to mete out the harshest penalties, and because no means of enforcement exists beyond these measures, the effectiveness of arbitration is lessened. The efficacy of arbitration judgments is ultimately limited by the DDC board's resolve.

It is legal for binding arbitration awards to be confirmed in civil court under New York law, which adds clout to the arbitration judgment and to require the person against whom the judgment was made to pay an additional 15 percent of the award for legal fees. This latter provision is rarely used, however, because the threat of its use is effective enough; besides the extra fees, public knowledge and damage to reputation would result. And reputation is the one currency that all traders work hard to maintain.

A diamond dealer went to the courts to have an arbitration judgment confirmed. When the judgment was confirmed, the person who hadn't complied with the judgment was served with papers, and he paid what he owed the family. Though the dealer was reluctant to use the courts in this limited way, he had noticed that the debtor was living extravagantly—obviously able to pay him what he owed. Among all his creditors, he paid only this family.

◆

A prime reason the club is loath to enforce the most severe arbitration penalties is illustrated by the Martin Rapaport case. Rapaport disproved Koskoff's statement that "Anyone who goes to civil court without permission of the club will suffer expulsion and blacklisting, which is likely to put a diamantaire [diamond trader] out of business throughout the diamond world" (1981:207–8).

Soon after Rapaport began his controversial price sheets in 1978, the DDC claimed he had violated a bylaw, maligning the integrity of certain members, and initiated an antitrust suit against him. He countersued, the FTC investigated the club's possible restraint of trade against Rapaport, and Rapaport was reinstated as a member after an undisclosed court settlement.

Thereafter, the DDC became extremely skittish about expelling members. As Bernstein noted, "The Rapaport controversy has made the club much more reluctant to expel members—it is concerned not only about the expelled member bringing suit, but it also fears that too many expulsions will revive the Federal Trade Commission's interest in its activities" (1992:156). Many DDC members agreed that the Rapaport battle was a turning point.

Before considering a DDC expulsion, the club now first seeks a court order that affirms its decision. Furthermore, club members are less inhibited about going to outside courts and going to arbitration armed with legal representation. These are serious departures. Though DDC members have always been allowed attorney representation, it is increasingly the standard rather than the exception.

The DDC shies from actions that might provoke lawsuits against the club. Since the amounts of money involved have grown exponentially and lawsuits are consequently expensive, the DDC has taken a defensive posture. Because the DDC board is less willing to enforce arbitration judgments, litigants are less concerned about the consequences. The decrease in enforcement undermines both the members' belief in arbitration and the authority of arbitration and the DDC. This trend ultimately means that the sense of community within the DDC and the communal power of social sanction are both lessening.

The presence of lawyers at arbitration reminds the participants of the court option. When lawyers are present, the chairman of the arbitration panel must achieve a balance between maintaining authority in the arbitration room and not overstepping procedural bounds that could jeopardize the ruling.

Many of my conversations with New York traders concerned how effective they considered arbitration to be, given these developments.

People don't want to be arbitrators anymore, one arbitrator lamented. They're shying away from responsibility, like the rest of society. It's hard to find bright and honest people, said another. Seriously, he added, someone with Alzheimer's was put on the ballot once and was reelected! Anyone can run, and some really bad people have run. Arbitrators are supposed to be neutral, but it's a very political process, he concluded.

One longtime arbitrator noted a person who allowed himself to be suspended from the club and then went to the New York court, hoping for a more favorable judgment. Though the court dismissed the case, the arbitrator was disturbed about the development. Arbitration was becoming more complicated in general, and litigants were using various delay tactics to escape negative judgments. If the judgments cannot be rendered in a timely fashion, he said, this alone denigrates the value of the process for recovering losses and affirming reputation among one's peers.

A subtle but seismic shift from compromise between the litigants, the traditional arbitration solution, to "settlement," the typical win-or-lose legal remedy, is dramatically altering the process. Yet many expressed their enduring belief in arbitration's ultimate viability regardless.

An arbitrator agreed that strict legal attention to procedural protocol was necessary. As a result, he had made simple changes. For example, he now allows only English to be spoken during arbitration, to avoid procedural problems that could occur if some of the participants spoke Yiddish.

Another arbitrator was critical of the DDC board for not backing all arbitration decisions—but 99 percent of the time the board of directors still backs us, he conceded. So such exceptions were rare. We get very upset with the board when it doesn't back our decisions, he said. Sometimes we have a discussion with the board and it's straightened out, he added.

A dealer, concerned that arbitration had less authority than it used to, noted that hard times had increased the number of cases. There were something like seventy cases in the works, he said, whereas years ago there were only about ten to fifteen cases a year. He thought fewer cases were settled in these difficult times. He also thought Bernstein's estimate that about 85 percent of cases are resolved in prearbitration procedures mandated by the DDC seemed high. During the prior year, he said, he and his firm had been involved in three cases—pretty unusual for their four-man operation. In one case, the opposing litigant was sentenced to prison for eighteen months, and they recovered perhaps 20 percent of their loss. That was a big deal. The other two cases are going nowhere, he said; in one case, he said, the lawyers are asking them to prove intent when it's clear that they gave a stone to someone on consignment, and he sold it for much less to someone else.

I was told about a dealer who went through the courts because he was frustrated by the lack of results in arbitration. The manufacturers who told me about this case explained that diamond club rules don't allow members much recourse. In this case, the courts were being used to speed a slow arbitration process—which was ironic, since court cases normally progress glacially. It's like a family dispute, they explained. The club tries to settle things so the disputants will stay in the club. Often that means that wrongs are not really redressed, which is why it's so tempting to go outside the DDC and to use a lawyer, they said.

If arbitration honors compromise, ultimately everyone and no one is satisfied.

An arbitrator was pleased about some of the positive changes he saw. Though he believed that arbitration worked because of speed, common sense, arbitrators' knowledge of the business, and confidentiality, he was also pleased that the DDC was now posting the names of litigants involved in upcoming arbitrations. Here was an instance of the system adapting well to changing times, he said. If Yusio has a claim against Joe, and Beryl wants to do business with Yusio, he can consult Joe about what's going on with Yusio, and find out for himself what was involved in the dispute. The arbitrator thought that this reform toward greater openness that still preserved the privacy of the process would benefit everyone.

Additionally, he wasn't bothered that lawyers were now present in the proceedings. Since the settlement process is supposed to precede formal arbitration, he said, it's not proper to continue to try to have the litigants settle the case once it's in formal arbitration. Because decisions have been vacated in the New York courts for that reason, they do not do it anymore. The courts have therefore been adhering to arbitration's true purpose. Arbitrators simply have to be more careful about procedural matters, he said. Other than being concerned that the DDC is less able to enforce its decisions, he felt that arbitration still worked well.

Controlling the lawyers during arbitration was an issue a few arbitrators noted. One of the most experienced and elderly arbitrators said he "had to keep the lawyers down" in his arbitrations. As chairman of his arbitration panel, he said, he had to acknowledge the lawyers, stay aware that they were there, and be mindful that they were watching procedure. An arbitrator must not be intimidated by their jargon, he added. He said he maintains order, keeps the lawyers firmly in place, and doesn't let the power of arbitration be diminished by their presence.

A manufacturer was similarly insistent and optimistic about maintaining the integrity of the arbitration process and philosophy in the face of changes, including the growing presence of lawyers. Tell me what your uncle said about me, he blustered from the start. That I'm old, right? Same as him! I was born in 1909. I'm an arbitrator, same as him—and for a long, *long* time. Arbitration hasn't changed, because the law is the same from the Talmud. Are you Jewish? Then you know. The law now is very strict. You can't have litigation without lawyers. We, the arbitrators, have to know a lot more about the law now. The litigant comes and sees that the other one is smarter than he is, so now they come with lawyers. But we don't let the lawyers abuse the arbitrators. So when we start, we say to the lawyer: We give you ten minutes. We are here without payment. We are elected. You are paid, so make it short and come to the point. Please do not interrupt. Interruptions will be $100. I'm very strict. Once I fined a lawyer. He had to pay it right away. It went to the club—never to us; it can't come to us. We don't allow a stenographer and tape recording. We also never make a judgment until all the evidence is in, and we don't rush with the judgment, because we want to find out everything. We have to take our time, get the whole story. The litigant says: I don't have it; I lost it. I tell him to find it. We have to have proof. Try to find it, we say. A signature proves everything. In court it's the same thing. Orally it's not good.

Another arbitrator said that though "lawyers try to throw their weight around," he manages. They use legal jargon, sure. But if he's the chair, he is not shy about saying things like, "I'm the boss here. This is my proceeding; I'm in charge. Do not use that language here. We will proceed in this way"—and so on. Still, he said, the arbitrators are always afraid they'll be taken to court, and the board of directors is likewise very careful. So things get "watered down." It can be very demoralizing because it's all done by volunteers, he added. You put in all this time, and you make enemies.

Some perceived preferential treatment in the system and said that each group favored its own. A considerable amount of negative feeling centered on the Hasidim. They were supposedly treated leniently in arbitrations because they had many children. One said the Indians never win and the Hasidim always win. Some thought Hasidim voted as a bloc for DDC board members and arbitrators. One elderly dealer was openly antagonistic in his remarks about the Hasidim.

"The biggest change in the industry I would say is the Hasidim, who have taken over the diamond end of the jewelry business. I'm trying to tell it to you without prejudice . . . I really am. The original Dutch Jews, we were known within the industry, a very honorable society. If you made a deal, you would shake hands with the person you're making the deal with, Jew and Gentile alike, and say *mazal* and *brucha*, and that was a contract. You could lose money, or make money, or anything else; that was a deal, and that's it. But the new group of people coming in, they don't seem to adhere to that at all—not at all, in my opinion. As I say, I'm trying not to be prejudiced. I *am* prejudiced. My experience has not been good."

He was angry about a case he'd been involved in. "It's very difficult. If you're not one of 'them'—that's what happened to us. They're so biased. We had a case against one of these Hasidim. *They suspended him for three months.* He literally stole $52,000 worth of diamonds from us—it was just pure theft. And we took him for charges in the club. And they suspended him for three months, you see. They said, 'Look, the man has to get a chance at making a living.'"

Whether the tensions involved in arbitration reflect the variety of ethnic and religious groups operating in the diamond industry today, it is easy and common to blame people who are not in one's own group for the ills in the system. Though arbitration is a system to resolve disputes, it necessarily and simultaneously reflects the divisions that exist among the groups that compose the membership.

A man in his fifties was critical of people reneging on their obligations and of a general decrease in trust in the business, all of which was reflected in arbitration in turn. His business had not done well, and his bitterness blazed through the blame. "The difference is that in our father's generation, the thought of going before an arbitration was the kiss of death to them," he insisted. "Now it's the reverse. It's not a big deal. They don't respect the arbitration. There are people who have no respect for the rulings. If Yankel gives me a stone on memo, I have no right to hand it over without telling him that I'm going to do anything. And the club can't decide on it. All of a sudden the club has a problem. You have certain kinds of dealings with traveling around the states and Fed Ex, and don't forget—when diamonds are given over, one member to another, you have no collateral. You hand over the diamond and he sells it; you have good faith. If he disappears, I lost my diamond, I lost my money, and *I* have to go prove it? There's always people who can't take it, and some of these people take off and hope for the best."

However, he appreciated how the club was responding. "The membership—the industry—is fighting. They are going to file criminal charges. It's a good thing [that the pictures of those called to arbitration are posted in the club]. They always had creditors' committees, but now they have more clout than 'Let's try to let the guy live.' No! What about all the people he took? The fact that he has ten children and a house and you listen to the whole story—well, excuse me, it's not the way you do business."

One dealer believed that some people deliberately abuse the system of trust that the diamond industry enshrines. Some people can be real scam artists, she said. One labeled "baloney" another's saying that the new people in the business don't have the same integrity as those who grew up in the business. Still others attributed the so-called degradation of the trust ethic to a human tendency to blame circumstance and others when times are tough and people feel vulnerable.

People told me they were displeased with delays and no-shows in arbitration. Several disliked extensions granted for flimsy excuses. One snapped: You couldn't check to make sure that the person actually did leave the country to go to his mother's funeral? People often cite similar circumstances in seeking to avoid attending the arbitration on the agreed-on date. Two arbitrators talking to each other noted that a person had not shown up for his arbitration. One of them said he should be fined for not showing up. But the other responded that everyone is afraid of lawsuits. They shook their heads gloomily and agreed: Things are really changing.

One longtime arbitrator noted the demoralizing effects of these more frequent delays. The process can take months or even a year, he said, from the time the complaint comes in until the matter is decided and appealed. Postponements happen a lot. You schedule the arbitration for one day next month, he said, and the day before, you call to confirm and find out the guy is going on a business trip that day. OK, next month—but that's no good because he won't be back yet. And then someone is sick—that kind of thing. In one way, that's good, he said, because people have incentive to settle in the meantime. Also, with the postponements it is OK to proceed if the person doesn't show up. That's in the rules. So that's good too.[5]

Another arbitrator had practical guidelines for dealing with such problems. He explained: We write a letter and tell him to come, and we do it three times; if he doesn't show up, then he's the guilty one. Simple. We have to give everybody time and a chance to show up. If he doesn't show up the third time and he doesn't write something, then he's the guilty one. Then, after ten days after the decision, he has a chance to make an appeal. He has to bring the money that he has to pay. He has to deposit it. If he's not guilty, he gets it back.

Various measures have sought to ensure fairness in arbitration; these include computer databases and the development of uniform principles that guide arbitrations worldwide. Despite such measures, however, I heard charges that some bourses are less fair than others. One person claimed that no non-Israeli has ever won an arbitration in Israel, and that the bourse there takes American passports to pressure Americans to stay for arbitrations. This person hoped that the WFDB would develop uniform arbitration rules for all the bourses, a difficult and laborious process. In order for this to happen, other bourses around the world would have to conform to the U.S. rules and constraints ensuring litigants' basic rights.

The perceived diminution of the power of arbitration was the clearest message I heard from a number of people, particularly the oldest members and the most longtime arbitrators. One of these men expressed disappointment about the dilution of authority that diminishes the effectiveness of arbitration.

---

[5] "In the event that a litigant fails to appear, refuses to proceed or unduly delays the proceedings in the judgment of that panel, the arbitration panel may proceed to a determination of the matter and make a decision and award" (DDC By-Laws, Article XII, Section 1e, 1999).

"We don't have as much power as we used to because of the litigious nature," he lamented. "People will take lawyers more than they used to twenty, thirty years ago. And the lawyers very often have sued the club because of technicalities, because of procedure, and the club has suffered very many financial losses because of that. Of course, 85 to 90 percent [of the decisions] go through. But some very important cases are held up because lawyers bring it to court. The club doesn't have the power to enforce the ruling that we've made. They're using the courts. In this respect it's changed the *character* [of arbitration]. We don't have the same power that we used to have. [In the past the arbitration decision] was much more ironclad. It was enforced. Much easier than today. They wouldn't go out[side the club], and they wouldn't hire lawyers, and they wouldn't go against the club. It's a new world. It's a more open society. People know more. It's like everything else. They're more aware of everything."

Many also noted consolidation within the trade. As a result of consolidations, the amounts of money involved in many cases are far larger than they used to be. A trade writer, while touting the benefits of arbitration, said lawyers are involved because the amounts in dispute are so much higher than they used to be. Years ago it was "one on one," but because the companies are bigger now, they need their lawyers as they proceed with arbitration.

Disputes still have to be solved in-house, said another dealer. A new arbitrator who believed in the process told me that he ran for the position because it's important for honest people to become involved. An ordained Orthodox rabbi in his eighties, Eli had been a diamond dealer and an arbitrator for many years and had similar views. It's like this, he said: the board of directors and the members of the DDC do not back the decisions of the arbitrators. He spread his big hands on the table between us and continued. It has to do with the fact that the amount of money involved has grown. These larger amounts provide more of an incentive to use lawyers, which provides more incentive to bring suits against the club, all of which makes the club more wary in general. There used to be an unwritten law that you couldn't go to court. That was it. Courts are now used, but courts usually turn the cases back to the DDC.

The work of an arbitrator is not popular, Eli said. Someone will always hate you and consider you unfair, and then after all that, your decisions don't have the clout they should have. He said he lost his best friend because someone deliberately gave him the wrong information about a case.

Traders' bottom-line assessments took practical accounting of the changes. Most seemed to feel that arbitration was a system that—with all its problems—still stood for the integrity of the business and was its symbol of trust. Arbitration is still essential to the business, concluded one, and that's that. Arbitration is an extremely positive part of the business.

One dealer lauded what she called the "courtly system of arbitration." She worried that emphasis on the problematic areas of arbitration exposes the structure of trust as essentially fragile. Trust is delicate, true, she said. Though she knew the flaws of arbitration and worried that people take advantage of the system, she was impressed by arbitration's overall elegance and efficacy. Another dealer summarized the business and arbitration this way:

I have a policy in life, she told me. When business is good in general, payment is easier; when business is hard, people are trying to survive and some of them get "off line" a little bit. The people I trade with are very moral. I work with a handshake, in Belgium, Germany—and here. The tendency in general is that it's not what it used to be because times are harder. I deal only with references. I need to have a background on a person unless the money is on the table. We are a big family. Yes, I call it a big family—there's always a brother, a father, an aunt; and at the end of the day, you know the good ones; you basically know the bad ones. And then at the end of the day, you have your surprises—you always have a surprise.

If somebody introduces me to somebody else—"This is my friend; give it to him"—my first question is: Who is [financially] guaranteeing? If he guarantees for him, then I'm basically all right, because if something goes wrong, then it's OK. We still don't work with contracts; we still work basically on a handshake. It protects me in a very small circle. If you work with out-of-town people, if you ship things, if you don't see the family, the store, you always take a risk. If you work locally in a small circle, it's OK. One of the first things I did was become a member of the club. It's the best way to meet people. It helps protect me. There's a board of directors; they do arbitration. If I have a problem with another member, I call it up to the board of directors and they set up the arbitration. I had to use arbitration once—slightly. I didn't have to do the arbitration ultimately. I called the executive office and asked with whom I could speak. Luckily, the person I talked to was the vice president. I explained, and he called the other person, and that straightened it out. This is why we like to be members. The more they see you, the more confident they are of you. It's all a matter of trust.

◆

As the diamond business has grown more complex, consolidated, and global, arbitration as a system of reconciling business disputes has likewise undergone change. But it is essentially the same system used through all of Jewish history throughout the world. Some of the strain that arbitration is under is due to the increased divisions within the community. Arbitration has traditionally acted as a way for the community to solve its problems internally, and this success in turn has enabled the community to define itself as separate and coherent. Over the years arbitration has done its job so well in the diamond business that it serves as its most revered symbol. But compromise is difficult, and genuine differences exist on 47th Street. Whether arbitration can continue to handle its conflicts as well as it has remains to be seen.

So far, the underlying values that inform the process of reconciliation have allowed arbitration to remain a workable vehicle to resolve conflict. The emphasis remains on settling, on keeping the disputes manageable, internal, and private, on recognizing the importance of maintaining reputation and relationship, and finally, on preserving the essence of community. These are the enduring values that keep arbitration central to the diamond business and emblematic of its finest qualities. It may be flawed, but it still works.

[9]

Conclusion

As we have seen in the preceding pages, diamond club business intertwines the social and the financial. Traders sit and mingle as they show goods, conversation in multiple languages is spirited and frequently heated, and news of business developments is interwoven with news of personal matters and current events.

This social foundation allows business to happen and makes business pleasurable. As the club is a public performance of human interaction containing vital information about the players, the connections help affirm who they are and that they are. Though different in various ways from relations in a Moroccan *suq*, interactions in the DDC marketplace are similar in principles. Reputation and prestige are at stake, as an individual diamond trader's honesty, expertise, and reliability are on daily display in this social arena. Private deals are made within the openness of the DDC setting. Arbitration in the DDC by the most respected of the traders' peers allows them—for the most part—to conciliate their conflicts internally and away from outside display.

The personal enjoyment of daily human interaction in general is enriched by such small gems of connection that we make continually. The quick check of each other's children with the bank teller, the exchange of empathic understanding about the heaviness of the day's heat, the acknowledgment of a return home by an acquaintance—these interactions smooth and add sheen to daily existence. We miss them when we don't have them. In diamond trading such interactions are more varied, more intense, and more frequent, with high pressure and steep risk added. At the beginning of the twenty-first century the challenges and pleasures of business endure. But the pleasure of the social—even the pleasure of making money—is not all that's going on here. It is the ongoing drama of new and old.

◆

In this book I have explored various aspects of the New York diamond business in order to illustrate how traders have been adapting to the changes of globalization at the end of the twentieth century and the beginning of the new millennium. While physical changes in the industry, including new sources of diamonds, new mechanisms for processing diamonds, and new uses for diamonds, continue to proliferate, the social changes have been by far the more monumental.

Marketing innovations, such as the Rapaport Diamond Price Sheet and the GIA grading standards, have helped streamline the trade and distribute product and pricing information widely. Dizzying improvements in electronic communications, including development of the personal computer, the fax, the cell phone, and the Internet, have tightened the trade by consolidating businesses, increasing competition, and driving many small traders out of business. This shift has discouraged many but has excited others.

A young man tried to describe the changes he had witnessed in the mere six years he had been working in his family's firm. The changes in the overall climate were dramatic, he told me in 1996. For one, the profitability of the business was less. People were more frantic. It was a smaller world than it used to be, with less time between transactions, and the orderly, more leisurely routines of the past were gone. He cited a simple example. He remembered when getting on a flight was a big deal. The flight would take sixteen hours. You had to take your time. You would go for a longer time because you had a lot to do, and the flights were few and far between. Now, though, with innovations like faxes and the Internet, life had become harder and faster. Business was much more competitive, with no breathing room.

People have maintained some of their old ways, however, while embracing the modern innovations in response to these gradual but monumental changes. They are selective, sometimes reluctant, sometimes eager. For example, they see that the computer can help them keep track of their inventory, and they experiment with other applications. Though many plunge in, others resist Internet trading, preferring the face-to-face contact for the reliability it helps provide. Traders' abilities are different, and though some of them are overcome by the obstacles, others see the obstacles as unique challenges that inspire them to gain an ultimate advantage. The Internet has expanded opportunities for creative entrepreneurs though risks are also high (Petzinger 1999). Women seem to have as much a chance as men at achieving success though some barriers in perception remain.

The men and women who contemplate entering the business are confronted with more career choices than ever, but the family firm still has its appeals. The American ethic of individualism rivals the pull of family holism—at least in some cases. Autonomy within the family may *reconcile* competing claims on a young person's loyalty. The family firm may harness these opposing forces in a particularly productive manner. Despite the problems attached to entering this perilous business at the beginning of the new millennium, many family firms are successful.

As a business magazine stated, "The family, it seems, embodies an unusual mixture of business interests and incentives. Its members trust one another. It cares about its fortune, but also about its good name; about the present, but also about posterity" ("In Praise of the Family Firm," 1996:16).

The men and women in these pages display a range of ways in which they have carved a niche for themselves within their firms. Though the risk is increasing, the impetus to be a differentiated individual within a family circle remains a powerful draw for those brave enough to stay within the business and to continue the line. Families and business might be good for each other.

◆

As they spoke with me about their involvement with the diamond business, the old traders asserted the continuing importance of their lives. While not explicitly intent on teaching the younger generation of traders, they knew that what they believed and practiced should be passed down. They were each living lives that served as examples. Each diamond trader who told me about working in old age seemed to recognize both the luck and the pluck in his own situation. Working in old age was a triumph that they seemed happy to display.

Some gerontologists contend that we burden the elderly with our focus on usefulness and productivity in the United States, and perhaps elderly diamond dealers reflect a cultural norm of productivity that traps them (for example, Moody 1993). But perhaps the old diamond dealers who continue to work would disagree. Most of them just like to keep working, and we should take them at their word, avoiding prescriptions about how to age, whether by activity *or* leisure.

At the DDC the younger traders perceive the old diamond dealers as the varied individuals they are. If they are talented, they are honored for these abilities, and when they fail in physical or cognitive health, their frailties are mourned. If they are able to continue working, the choice to work seems to rest mostly with them. Those who are gone cannot be replaced. The histories of people trail them. And so, they remain dealers, manufacturers, brokers, arbitrators—role models for the others who, whether they know it, are learning from them.

◆

Moore has written, "In the life-term social arena, one who is neither expelled nor obliged to leave must live constantly with his life history. One cannot avoid having been at one time or another party to taunting and punishing someone. It is not possible to reach maturity without having ill will toward someone. One cannot avoid taking sides in the disputes of others even if lucky enough not to be directly involved. But if the disputes are one's own, cumulation may be of bad relations as well as good ones" (1978:71–72).

Disputes are solved in some cases and not in others as arbitration succeeds and also falls short in these more complicated days. Though compromise is usually attained, traders also resort to outside courts, a practice increasingly threatening to the viability of the group and to its enshrined method of resolution. Arbitration, that elegant system for the mediation of industry conflict, reflects the strains within the trade and the pressures from the outside that weaken it. Whether the trade can uphold arbitration's noble traditions as the twenty-first century proceeds will ultimately be up to the resolve of its members—and its officials—to adamantly honor and to insist on preserving its cherished principles of fairness and integrity.

◆

The most dramatic factor providing the backdrop for changes in buying and selling in the New York diamond business is the worldwide global economy. An individual's judgment and experience may not be sufficient for her or him to survive as the diamond trade becomes increasingly global and compressed. On the other hand, many are in a good position to modify niches that they have taken years to carve out. The most savvy small-business people are still able to create the personal niches that many customers continue to want—and others may be unable to compete with these specialists.

Though the amazing, intense speed of information exchange both reflects and causes greater interweaving of the world's economies, the diamond industry has *always* been a worldwide business known for its transnational character. The earliest diamond trade routes were global trails. De Beers and the CSO acted globally from the start, and the effects of a new find in Brazil, Botswana, or Russia were always reflected in a CSO price change that would in turn affect the dealers and brokers on the street in New York, Bombay, and Antwerp. How De Beers and Namibia or Australia negotiated had an almost immediate effect on prices of rough everywhere—and still does. The rest of the world is catching up. Each country's economy is becoming more interdependent with the others, and this "global contagion" produces complex waves of immediate impact, the repercussions of which are hard or impossible to predict (Kristof and WuDunn 1999).

Diamond traders have operated as members of the international umbrella organization of the WFDB, acting locally and thinking and being affected globally. Whether trading stones or arbitrating disputes, they have operated in an international enterprise. Members always have had to act within the laws of the locales in which they were situated, but they were also part of the worldwide WFDB. The ancient, global roots and routes of the business have made diamond traders well conditioned for the changes that are now affecting the entire planet. Perhaps more than people in most other businesses, they understand how business cycles, international relations, and the world economic situation affect the small trader. The oldest ones have the longest perspective about business cycles over time. I discussed the cycles with three men in the DDC who were lamenting the current state of the business.

The two oldest men were eighty-nine and eighty-four; the third was in his seventies. The three agreed: Everything is tight now. It's difficult to get goods, profits are shrinking, and businesses are going under. The two oldest men reminisced about worse times and commented that times always seem bad. They ticked off the dates of some of the particularly bad years and nodded gravely with each memory.

The recession and high interest rates of the early 1990s produced a credit crunch. The Asian economic crisis in the late 1990s and the reduction in the role of the middleman sobered the consumer market for luxury items and reduced profits overall. Yet by 1999, the Asian woes were receding, and some hoped for a strong resurgence. The American economy was enjoying an unprecedented period of prosperity until shortly before the George W. Bush presidency, when a new downturn began.

More examples of economic and political interaction were dramatically evident in 2001, especially the tremors of uncertainty that followed the terrorist attacks on the United States on September 11, 2001. The international diamond industry was again on center stage as news spread about the use of pilfered diamonds to finance civil wars and to deplete national economies. In response, the newly formed World Diamond Council urged the U.S. Congress to ban the import of gems from conflict-riddled regions of Africa and to support a certification-of-origin system with a global database for packets of approved diamonds (Cowell 2001). By taking such a proactive—and unprecedented—stance against diamonds linked to civil wars, the worldwide diamond industry was seeking to head off an international boycott that could destroy the trade like a pin that pricks a fancy balloon.

At the same time, De Beers announced that it was entering the polished diamond and jewelry market through a $400 million partnership with luxury retailer LVMH Moet Hennessy Louis Vuitton (Cauvin 2001). These jewels would be exclusively branded with the De Beers name. Although some considered it a risk to magnify the De Beers name at a time when diamonds were being linked with war and brutality, De Beers was shielding itself by attaching to a well-known luxury retailer. Furthermore, the potential benefit of gaining direct access to the United States, the largest consumer of diamond jewelry, without provoking U.S. antitrust charges was for the first time possible through this partnership. Branding its own diamonds to hinder diamonds linked to the civil wars from reaching the market might also be a sly way for De Beers to deal with its shrinking hold on the world's supply and to polish its image, according to an analysis by Rapaport (2001).

In addition, De Beers was disentangling itself from Anglo-American P.L.C., the largest mining company in the world, in order to bolster its stock and to allow the Oppenheimer family a way to regain control of the diamond business. Perhaps the partnership was also a way for De Beers to deal with its history as a target of antitrust suits (Cowell and Swarns 2001). "Nicky really knows what he's doing," commented Martin Rapaport in a mixed-praise assessment of the "brilliant" plan by Ernest Oppenheimer's grandson (Rapaport 2001).

New York traders, affected by these changes more than ever, have continued to give a mixed assessment about the future. "I don't see the prognosis as very good for dealers," a past president of the DDC, William Goldberg, was quoted as saying by a trade magazine in mid-2001. He continued, "I say to the younger generation, 'Get involved with computers or accounting. You need to reschool yourself and find another way of making a living.' I cry when I say it. I love the people here. But we have to face reality." Yet, Goldberg—and other businessmen intent on scoping out the not-so-evident benefits—also held out hopes that the De Beers plan to partner with LVMH would benefit traders by strengthening the diamond trade and increasing the market as a whole (Bates 2001).

These new and dramatic steps attest to the way attention can be focused on an issue quickly and effectively in an Internet-linked world. Much of the speedy response to the concern about diamonds linked to the civil wars can be attributed to the efforts of the nongovernmental organization Global Witness. That this tiny English organization was able to provoke the huge diamond trade, the United Nations, and the United States to act to stem this tainted aspect of the trade testifies to the intensity of our connectedness. The trade understands how susceptible it is to the whims of the world and how quickly the allure of diamonds can be transformed into repugnance by world perception of their links to brutality. Never mind that only the tiniest percentage of diamonds is from conflict-riddled areas—and that numerous legitimate economies and the trade support millions of people. The industry knows that it must act ever more quickly and definitively. Its formation of the World Diamond Council to responsibly and forcefully respond to world events reflects this understanding. Time will tell. . . .

The diamond's essential identity, as an expensive luxury item purchased with discretionary money, makes it fragile in soft markets and hard times. Though East Asia recovered fairly well from its "Asian flu" of the 1990s, its problems reminded everyone of the vulnerability of the enterprise. The terrorist attacks of September 11, 2001, reinforced the message that diamonds are not necessities. At the same time, some were inspired to make long-delayed commitments, capped with diamond rings. Alongside these fluctuations, New York seems likely to retain its reputation as the premier center for the largest and finest diamonds. This niche relies on the expertise of the gifted cutters who transform the rough into polished. More manufacturers and traders may go into producing and designing jewelry to capitalize on this niche. The focus on large diamonds and on their subsequent transformation into unique pieces of jewelry provides a countervailing force to commodification.

Middlemen may continue to be squeezed out, but new possibilities for talented entrepreneurs also exist. They may create new and narrower niches as other inventive individuals are doing in other businesses. As New York adjusts to the new conditions of the world economy, it may continue to thrive in its latest response to the pressures of change. Will and grit will be necessary. How the small traders on 47th Street will ultimately fare will hinge on their individual determination and resourcefulness as well as on the tumult of emerging events in the world.

Mazal.

Bibliography

Abu-Lughod, Lila. 1991. "Writing against Culture." In *Recapturing Anthropology*, edited by Richard G. Fox. Santa Fe: School of American Research Press.

American Diamond Industry Association, Inc. (ADIA) newsletters. Published 1982–1997. New York.

"Angola: Government Okays Rebels' Talking to Mining Firms." 1997. *Diamond World Review* 96 (March–April): 11–12.

Appadurai, Arjun, ed. 1986. "Introduction: Commodities and the Politics of Value." In *The Social Life of Things: Commodities in Cultural Perspective*. New York: Cambridge University Press.

———. 1996. *Modernity at Large: Cultural Dimensions of Globalization*. Minneapolis: University of Minnesota Press.

"Argyle Executive Sees End of Single Channel." 1997. *Diamond World Review* 95 (December 1996–January 1997): 34.

Balfour, Ian. 1992. Introduction, *Famous Diamonds*. 2d ed. Essex (UK): NAG Press.

Barth, Michael C., William McNaught, and Philip Rizzi. 1995. "Older Americans as Workers." In *Older and Active: How Americans Over 55 Are Contributing to Society*, edited by Scott A. Bass. New Haven: Yale University Press.

Bates, Rob. 1995. "Judged on Their Merit." *National Jeweler*. March 16, 1995.

———. 2001. "Panel to Dealers: Find Another Career." *Jewelers' Circular Keystone*. May 1, 2001.

Becker, Howard P. 1950. *Through Values to Social Interpretation*. Durham: Duke University Press

Behar, Ruth. 1995. "Writing in My Father's Name." In *Women Writing Culture*, edited by Ruth Behar and Deborah A. Gordon. Berkeley: University of California Press.

Belcove-Shalin, Janet S. 1988a. "Becoming More of an Eskimo: Fieldwork among the Hasidim of Boro Park." In *Between Two Worlds: Ethnographic Essays on American Jewry*, edited by Jack Kugelmass. Ithaca: Cornell University Press.

——. 1988b. "The Hasidim of North America: A Review of the Literature." In *Persistence and Flexibility: Anthropological Perspectives on the American Jewish Experience*, edited by Walter P. Zenner. Albany: State University of New York Press.

Belcove-Shalin, Janet S., ed. 1995. *New World Hasidism: Ethnographic Studies of Hasidic Jews in America*. Albany: State University of New York Press.

"Beneath Frozen, Desolate Terrain Lies a World-Class Diamond Mine." *Providence Journal*. June 21, 1998: A25.

Benedetti, Laura Robin, Jeffrey H. Nguyen, Wendell A. Caldwell, Hongjian Liu, Michael Kruger, and Raymond Jeanloz. 1999. "Dissociation of CH4 at High Pressures and Temperatures: Diamond Formation in Giant Planet Interiors?" *Science* 286, no. 5437 (October 1): 100–102.

Bernstein, Lisa. 1992. "Opting Out of the Legal System: Extralegal Contractual Relations in the Diamond Industry." *Journal of Legal Studies* 21: 115–57.

Berquem, V. 1988. "Bourses More Than a Place to Sell." *Jewellery News Asia*. (August).

Blauer, Ettagale. 2001. "Still on the Cutting Edge." *New York Diamonds* 63 (March): 38–48.

Bonacich, Edna. 1973. "Theory of Middleman Minorities." *American Sociological Review* 38: 583–94.

Bonner, Raymond. 1999. "U.S. May Try to Curb Diamond Trade That Fuels Africa Wars." *New York Times*. August 8, 1999: A3.

Bourdieu, Pierre. 1990. *The Logic of Practice*, translated by R. Nice. Stanford: Stanford University Press.

Boyarin, Jonathan. 1988. "Waiting for a Jew: Marginal Redemption at the Eighth Street Shul." In *Between Two Worlds: Ethnographic Essays on American Jewry*, edited by Jack Kugelmass. Ithaca: Cornell University Press.

——. 1992. *Storm from Paradise: The Politics of Jewish Memory*. Minneapolis: University of Minnesota Press.

Branch, Hilda. 1998. "Verona–Brussels." In *Hitler's Exiles: Personal Stories of the Flight from Nazi Germany to America*, edited by Mark M. Anderson. New York: W.W. Norton.

Brooks, André. 1992. "Behind the Glitter, Recession Hurts an Exotic Bazaar." *New York Times*. March 29, 1992: section 10, p. 13.

Callahan, Maximillian S. 1996. *Insider Secrets to Diamond Dealing: How Real Money Is Made*. Boulder, Colo.: Paladin Press.

Cauvin, Henri E. 2001. "De Beers and LVMH Linking Up in Retail Venture." *New York Times*. January 17, 2001: W1.

Charny, David. 1990. "Nonlegal Sanctions in Commercial Relationships." *Harvard Law Review* 104, no. 2: 375–467.

Cohen, Abner. 1974. *Two-Dimensional Man: An Essay on the Anthropology of Power and Symbolism in Complex Society*. Berkeley: University of California Press.

——. 1981. *The Politics of Elite Culture: Explorations in the Dramaturgy of Power in a Modern African Society*. Berkeley: University of California Press.

Cole, Ernest. 1967. *House of Bondage*. New York: Random House.

Cole, Thomas. 1992. *The Journey of Life: A Cultural History of Aging in America*. Cambridge: Harvard University Press.

Collins, Alan T. 1998. "Diamonds in Modern Technology: Synthesis and Applications."

In *The Nature of Diamonds*, edited by George E. Harlow. Cambridge (UK): Cambridge University Press (in conjunction with the American Museum of Natural History).

Collins, Wilkie. 1994 [1868]. *The Moonstone*. In *Wilkie Collins: Three Great Novels*. New York: Oxford University Press.

Conflict Diamonds: Possibilities for the Identification, Certification and Control of Diamonds. 2000. Briefing document. London: Global Witness.

Contreras, Joseph. 1992. "The Lure of Diamonds." *Newsweek* 120, no. 19 (November 9): 34–35.

Cowell, Alan. 2001. "Congress Asked to Help Police the Rogue Diamond Trade." *New York Times*. January 19, 2001: W1.

Cowell, Alan and Rachel Swarns. 2001. "Disentangling a Worldwide Web of Riches." *New York Times*. February 2, 2001: C1, 3.

Crossette, Barbara. 2000a. "South Africa to Ask for International Gem Certification." *New York Times*. November 30, 2000: A10.

———. 2000b. "U.N. Confirms Liberia's Role in Smuggling of Diamonds." *New York Times*. December 20, 2000: A7.

———. 2000c. "Gem Sanctions Sought by U.N. Are Delayed." *New York Times*. December 21, 2000: A18.

Daley, Suzanne. 1998. "Angolan Rebels Still Break Diamond Embargo, Rights Group Says." *New York Times*. December 15, 1998: A7.

Davidman, Lynn and Shelly Tenenbaum. 1994. "Toward a Feminist Sociology of American Jews." In *Feminist Perspectives on Jewish Studies*, edited by Davidman and Tenenbaum. New Haven: Yale University Press.

Davis, John. 1996. "An Anthropologist's View of Exchange." *Social Anthropology* 4, no. 3: 213–26.

DDC Arbitration By-Laws. 1993. New York: Diamond Dealers Club, Inc.

DDC By-Laws. 1999 and 1980. New York: Diamond Dealers Club, Inc.

"DDC Efforts Seek Higher Profits for Members." 1995. *New York Diamonds* 32: 81–82.

"The DDC: Gateway to the Diamond Heartland." 1997. http://www.Diamonds.net. September 5, 1997.

"DDC President Advises Members to Specialize." 1995. *New York Diamonds* 29 (March): 78–81.

"De Beers Ending Agreement to Sell Russian Diamonds." 1997. *New York Times*. January 3, 1997: D2.

"De Beers Mine to Keep Botswana as Diamond Leader." 1999. http://www.Diamonds.net. September 2, 1999.

"De Beers Pushes Millennium-Themed Diamonds." 1999. http://www.Diamonds.net. August 13, 1999.

"De Beers Testifies at Congress Hearing on Conflict Diamonds." 2000. http://www.Diamonds.net. May 16, 2000.

"Diamonds.com. 2000." http://www.Diamonds.net. May 30, 2000.

"Diamond Dipping." 1999. http://www.Diamonds.net. March 31, 1999.

"Diamond Trade and Industry." 1994. In *Encyclopedia Judaica* 6. New York: Macmillan Co.

Diamonds: A Cartel and Its Future. 1992. London: Economist Intelligence Unit. Special Report no. M702.

"Diamonds: The Cartel Lives to Face Another Threat." 1987. *The Economist* 302 (January 10): 58–60.

"Diamonds on the Internet." 1995. *New York Diamonds* 33: 47–57.

Dominguez, Virginia R. 1993. "Questioning Jews." *American Ethnologist* 20, no. 3: 618–24.

Durkheim, Emile. 1933 [1964]. *The Division of Labor in Society.* New York: Macmillan Co.

El-Or, Tamar. 1994. *Educated and Ignorant: Ultraorthodox Jewish Women and Their World*, translated from Hebrew by Haim Watzman. Boulder, Colo.: Lynne Rienner Publishers.

Epstein, Edward J. 1981. *The Diamond Invention.* New York: Simon & Schuster.

Fabian, Johannes. 1983. *Time and the Other: How Anthropology Makes Its Object.* New York: Columbia University Press.

Farber, Zvi. 1996. "Arbitration: DDC Problem Solver." *New York Diamonds* 37 (October/November): 62–64.

Featherstone, Mike. 1995. *Undoing Culture: Globalization, Postmodernism and Identity.* Thousand Oaks, Calif.: Sage Publications.

Federman, David. 1985. "Diamonds and the Holocaust." *Modern Jeweler*. 39–46, 72.

Fisher, Ian. 1998. "Where War Is Forever, the Diamonds Are Cheap." *New York Times.* December 25, 1998: A4.

Frantz, Douglas. 1994a. "How G.E. Plays for Keeps in Diamonds." *New York Times.* September 18, 1994: Business section, p. 1.

——. 1994b. "G. E. Wins in Diamond Price Case." *New York Times.* December 6, 1994: D2.

Freedman, Samuel G. 2000. *Jew vs. Jew: The Struggle for the Soul of American Jewry.* New York: Simon & Schuster.

French, Howard W. 1995. "In Sierra Leone, Darkness, Not Diamonds' Dazzle." *New York Times.* October 9, 1995: A4.

Friedman, Jane. 1979. "Women: New Facet of the Diamond Trade." *New York Times.* February 18, 1979: F2.

Fritsch, Emmanuel. 1998. "The Nature of Color in Diamonds." In *The Nature of Diamonds*, edited by George E. Harlow. Cambridge (UK): Cambridge University Press (in conjunction with the American Museum of Natural History).

Furman, Miriam. 1996. "Women Can Succeed in Industry, Says DDC's First Female Director." *New York Diamonds* 34: 78–80.

Gaouette, Nicole. 2001. "Israel's Diamond Dealers Tremble." *The Christian Science Monitor.* June 1, 2001: 6–7.

Geertz, Clifford. 1979. "Suq: The Bazaar Economy in Sefrou." In *Meaning and Order in Moroccan Society: Three Essays in Cultural Analysis*, by Clifford Geertz, Hildred Geertz, and Lawrence Rosen. New York: Cambridge University Press.

"GIA Calls for Facts on Purported Diamond Treatment." 1999. http://www.Diamonds.net. March 23, 1999.

"GIA Sets to Work on Cut Question." 1999. http://www.Diamonds.net. January 12, 1999.

GIA website. http://www.GIA.org. December 4, 1999.

Giddens, Anthony. 1976. *New Rules of Sociological Method: A Positive Critique of Interpretative Sociologies.* London: Hutchinson.

Goldstein, Israel. 1981. *Jewish Justice and Conciliation: History of the Jewish Concili-ation Board of America, 1930–1968*. New York: Ktav Publishing House.

Gramsci, A. 1971. *Selections from the Prison Notebooks*, translated and edited by Q. Hoare and G. N. Smith. London: Lawrence & Wishart.

Green, Timothy. 1981. *The World of Diamonds*. New York: William Morrow and Com-pany, Inc.

Gregory, Sir Theodore. 1962. *Ernest Oppenheimer and the Economic Development of Southern Africa*. Cape Town: Oxford University Press.

Gross, Nachum, ed. 1975. *Economic History of the Jews*. New York: Schocken.

Gupta, Akhil and James Ferguson, eds. 1997. "Discipline and Practice." In *Anthropo-logical Locations: Boundaries and Grounds of a Field Science*. Berkeley: University of California Press.

Gutwirth, Jacques. 1968. "Antwerp Jewry Today." *Jewish Journal of Sociology* 10: 121–37.

Hahn, Emily. 1956. *Diamond*. New York: Doubleday.

Harden, Blaine. 2000. "Two African Nations Said to Break U.N. Diamond Embargo." *New York Times*. August 1, 2000: A3.

Harlow, George E. 1998a. "Following the History of Diamonds." In *The Nature of Di-amonds*, edited by George E. Harlow. Cambridge (UK): Cambridge University Press (in conjunction with the American Museum of Natural History).

———. 1998b. "From the Earth to Fashioned Objects: Processing Diamond." In *The Nature of Diamonds*, edited by George E. Harlow. Cambridge (UK): Cambridge University Press (in conjunction with the American Museum of Natural History).

———. 1998c. "Diamonds in the Twentieth Century." In *The Nature of Diamonds*. Cam-bridge (UK): Cambridge University Press (in conjunction with the American Mu-seum of Natural History).

Harlow, George E., Vladislav S. Shatsky, and Nikolai V. Sobolev. 1998. "Natural Sources of Diamond Other Than the Earth's Mantle." In *The Nature of Diamonds*. Cambridge (UK): Cambridge University Press (in conjunction with the American Museum of Natural History).

Hart, Matthew. 1999. "Where Thievery Is an Art." http://www.Diamonds.net. April 5, 1999.

———. 2001. *Diamond: A Journey to the Heart of an Obsession*. New York: Walker & Company.

Hazen, Robert M. 1999. *The Diamond Makers*. New York: Cambridge University Press.

Heilman, Samuel C. 1982. "The Sociology of American Jewry: The Last Ten Years." *Annual Review of Sociology* 8: 135–60.

Heilman, Samuel C. and Steven M. Cohen. 1989. *Cosmopolitans & Parochials: Mod-ern Orthodox Jews in America*. Chicago: University of Chicago Press.

Helmreich, William B. 1992. *Against All Odds: Holocaust Survivors and the Successful Lives They Made in America*. New York: Simon & Schuster.

Howe, Irving. 1976. *World of Our Fathers*. New York: Harcourt Brace Jovanovich.

Hyman, Paula E. 1994. "Feminist Studies and Modern Jewish History." In *Feminist Perspectives on Jewish Studies*, edited by Lynn Davidman and Shelly Tenenbaum. New Haven: Yale University Press.

Ilani, Zvi, Yitzhak Goldberg, and Jaakov Weinberger. 1991. *Diamonds and Gemstones in Judaica: A Selective Anthology from the Scriptures, the Mishna, the Talmud, Midrashim, and Responsa Literature.* Ramat Gan: Bar Ilan University.

"In Praise of the Family Firm." 1996. *The Economist* 338, no. 7956 (March 9, 1996): 16.

Jackson, Michael, ed. 1996. Introduction, *Things As They Are: New Directions in Phenomenological Anthropology.* Bloomington: Indiana University Press.

Jacobs and Chatrian. 1880. *Monographie du Diamant.* Antwerp and Paris.

"Japan Inching Back to Diamonds." 1999. http://www.Diamonds.net. April 21, 1999.

"Japanese Market Expecting Big Changes in 1997." 1997. *Diamond World Review* 96 (March–April): 32.

Jessup, Edward. 1979. *Ernest Oppenheimer: A Study in Power.* London: Rex Collings.

Joselit, Jenna Weissman. 1990. *New York's Jewish Jews: The Orthodox Community in the Interwar Years.* Bloomington: Indiana University Press.

Kahan, Arcadius. 1986. *Essays in Jewish Social and Economic History*, edited by Roger Weiss. Chicago: University of Chicago Press.

Kahn, Joseph. 2000. "World Bank Blames Diamonds and Drugs for Many Wars." *New York Times.* June 16, 2000: A12.

Kanfer, Stefan. 1993a. "A Cartel More Durable Than Diamonds." *New York Times.* September 26, 1993: 3,11.

———. 1993b. *The Last Empire: De Beers, Diamonds, and the World.* New York: Farrar Straus Giroux.

Keller, Bill. 1992. "De Beers May Be Losing Grip on Diamond Market." *New York Times.* September 2, 1992: A1.

Kertzer, David I. 2001. *The Popes against the Jews: The Vatican's Role in the Rise of Modern Anti-Semitism.* New York: Knopf.

Kirkley, Melissa B. 1998. "The Origin of Diamonds: Earth Processes." In *The Nature of Diamonds*, edited by George E. Harlow. Cambridge (UK): Cambridge University Press (in conjunction with the American Museum of Natural History).

Kirshenblatt-Gimblett, Barbara. 1987. "The Folk Culture of Jewish Immigrant Communities." In *The Jews of North America*, edited by Moses Rischin. Detroit: Wayne State University Press.

Knauft, Bruce M. 1996. *Genealogies for the Present in Cultural Anthropology.* New York: Routledge.

Knight, John and Heather Stevenson. 1986. "The Williamson Diamond Mine, De Beers, and the Colonial Office: A Case-Study of the Quest for Control." *Journal of Modern African Studies* 24, no. 3: 423–45.

Kockelbergh, Iris, Eddy Vleeschdrager, and Jan Walgrave. 1992. *The Brilliant Story of Antwerp Diamonds.* Antwerp: MIM, n.v. Ortelus Series.

Kopytoff, Igor. 1986. "The Cultural Biography of Things: Commoditization as Process." In *The Social Life of Things*, edited by Arjun Appadurai. New York: Cambridge University Press.

Koskoff, David E. 1981. *The Diamond World.* New York: Harper and Row.

Krajick, Kevin. 2001. *Barren Lands: An Epic Search for Diamonds in the North American Arctic.* New York: Henry Holt and Company.

Kranzler, George. 1995. "The Economic Revitalization of the Hasidic Community of Williamsburg." In *New World Hasidim: Ethnographic Studies of Hasidic Jews in America*, edited by Janet S. Belcove-Shalin. Albany: State University of New York Press.

Kristof, Nicholas D. and Sheryl WuDunn. 1999. "Of World Markets, None an Island." *New York Times*. February 17, 1999: A1.

Kugelmass, Jack. 1986. *The Miracle of Intervale Avenue: The Story of a Jewish Congregation in the South Bronx*. New York: Schocken Books.

Kugelmass, Jack, ed. 1988. Introduction, *Between Two Worlds: Ethnographic Essays on American Jewry*. Ithaca: Cornell University Press.

Laan, H. L. van der. 1965. *The Sierra Leone Diamonds: An Economic Study Covering the Years 1952–1961*. London: Oxford University Press.

——. 1975. *The Lebanese Traders in Sierra Leone*. The Hague: Mouton.

"Labs." 1999. http://www.Diamonds.net. February 15, 1999.

Lawlor, Julia. 1998. "Therapy for the Family Business." *New York Times*. November 22, 1998: Business section, p. 10.

Leary, Warren E. 1999. "Prospecting for Diamonds on the Outer Planets." *New York Times*. October 5, 1999: D3.

Legrand, Jacques (and others). 1980. *Diamonds: Myth, Magic, and Reality*. New York: Crown Publishers.

Leichter, Hope J. and William E. Mitchell. 1978. *Kinship and Casework: Family Networks and Social Intervention*. Enlarged edition. New York: Teachers College Press.

Levinson, Alfred A. 1998. "Diamond Sources and Their Discovery." In *The Nature of Diamonds*, edited by George E. Harlow. Cambridge (UK): Cambridge University Press (in conjunction with the American Museum of Natural History).

Lévi-Strauss, Claude. 1949. *The Elementary Structures of Kinship*. Boston: Beacon Press.

Liberman, Kopel. 1935. *L'industrie et le Commerce Diamantaires Belges*. Brussels: E. Duchatel.

Lipschitz, Simone. 1990. *De Amsterdamse Diamantbeurs [The Amsterdam Diamond Exchange]*. Amsterdam: Stadsuitgerverij Amsterdam.

Loeb, Laurence D. 1977. *Outcaste: Jewish Life in Southern Iran*. New York: Gordon and Breach.

Lopate, Phillip. 2001. "Fish Tale: Falling for a Live One." *New York Times*. January 5, 2001: B37.

Lubin, Albert J. 1982. *Diamond Dealers Club: A 50-Year History*. New York: Diamond Dealers Club.

Macaulay, Stewart. 1963. "Non-Contractual Relations in Business: A Preliminary Study." *American Sociological Review* 28: 55–69.

MacLeod, Scott. 1992. "Diamonds Aren't Forever." *Time*. October 12, 1992: 73.

Malinowski, Bronislaw. 1922. *Argonauts of the Western Pacific*. London: Routledge.

McDonald, Hamish. 1993. "Diamond Industry: All That Glitters." *Far Eastern Economic Review* 156, no. 32 (August 12, 1993): 74.

McNeil, Donald G. Jr. 1998a. "Find a Diamond in the Sand? Just Don't Pick It Up." *New York Times*. April 27, 1998: A4.

——. 1998b. "South African Industrial Giant Moving to London." *New York Times*. October 16, 1998: C1.

——. 1999. "A Diamond Cartel May Be Forever." *New York Times*. January 12, 1999: C1.

Melcher, Richard A. 1992. "Can De Beers Hold on to Its Hammerlock?" *Business Week*. September 21, 1992: 45–46.

Michelle, Amber. 1997. "The Many Faces of 47th Street." http://www.Diamonds.net. September 5, 1997.

Mintz, Jerome R. 1992. *Hasidic People: A Place in the New World*. Cambridge: Harvard University Press.

Mitchell, William E. 1978. *Mishpokhe: A Study of New York City Jewish Family Clubs*. New York: Mouton Publishers.

———. 1988. "A Goy in the Ghetto: Gentile-Jewish Communication in Fieldwork Research." In *Between Two Worlds: Ethnographic Essays on American Jewry*, edited by Jack Kugelmass. Ithaca: Cornell University Press, 1988.

Molho, Rene. 1994. *They Say Diamonds Don't Burn: The Holocaust Experiences of Rene Molho of Salonika, Greece*. Berkeley, Calif.: Judah L. Magnes Museum.

Moody, Harry R. 1993. "Age, Productivity, and Transcendence." In *Older and Active: How Americans Over 55 Are Contributing to Society*, edited by Scott A. Bass. New Haven: Yale University Press.

Moore, Sally Falk. 1978. "Old Age in a Life-Term Social Arena: Some Chagga of Kilimanjaro in 1974." In *Life's Career-Aging: Cultural Variations on Growing Old*, edited by Barbara G. Myerhoff and Andrei Simic. Beverly Hills: Sage Publications.

Morawska, Ewa. 1996. *Insecure Prosperity: Small-Town Jews in Industrial America, 1890–1940*. Princeton: Princeton University Press.

Moyar, A. 1960. *The Diamond Industry in 1958–1959*. Antwerp: Vlaams Economisch Verbond.

Myerhoff, Barbara. 1979. *Number Our Days*. New York: E. P. Dutton.

Myerson, Allen R. 1992. "A District That Was Not Forever." *New York Times*. September 4, 1992: D1.

Narayan, Kirin. 1993. "How Native Is a 'Native' Anthropologist?" *American Anthropologist* 95: 671.

Nessman, Ravi. 2000. "Diamonds Lift Botswana Out of Poverty." *Providence Journal*. November 12, 2000: D14.

Nestlebaum, Karen. 1996. "Market Week Brings Surge of Business to Club." *Rapaport Diamond Report*. October 11, 1996: 15.

———. 1999. "Industry Speaks Out to Uphold Ethics As Pegasus Invasion Nears." http://www.Diamonds.net. March 31, 1999.

———. 2001. "Main Street Retailers Tentatively Embrace the Web." http://www.Diamonds.net. January 30, 2001.

Newbury, Colin. 1989. *The Diamond Ring: Business, Politics, and Precious Stones in South Africa, 1867–1947*. New York: Oxford University Press.

Oliver, Roland. 1990. "The Unstoppable Rise of De Beers." *Times Literary Supplement* 4533 (February 16–22): 174.

Onishi, Norimitsu. 2000. "In Ruined Liberia, Its Despoiler Sits Pretty." *New York Times*. December 7, 2000: A1.

———. 2001. "Africa Diamond Hub Defies Smuggling Rules." *New York Times*. January 2, 2001: A1.

Parker, Daniel. "New York Competing on the Home Front." 1998. http://www.Diamonds.net. August 3, 1998.

Passaro, Joanne. 1997. "You Can't Take the Subway to the Field!: 'Village' Epistemologies in the Global Village." In *Anthropological Locations: Boundaries and Grounds of a Field Science*, edited by Akhil Gupta and James Ferguson. Berkeley: University of California Press.

Patwa, Subhadra. 1989. *Role of Women in Diamond Trade and Industry*. Bombay: Research Centre for Women's Studies, SNDT Women's University. A study sponsored by Industrial Development Bank of India.

Petzinger, Tom. 1999. *The New Pioneers: The Men and Women Who Are Transforming the Workplace and Marketplace*. New York: Simon and Schuster.

Putnam, Robert D. 2000. *Bowling Alone: The Collapse and Revival of American Community*. New York: Simon and Schuster.

"Rapaport, De Beers Deny Buyout Rumor." 1997. *Diamond World Review* 96 (March–April): 14–16.

"Rapaport Fancy Prices Provoke Protests." 1996. *New York Diamonds* 36 (August/September): 36–38.

Rapaport, Martin. 1995. "Discounting." http://www.Diamonds.net. February 3, 1995.

——. 1999. "Bottoms Up." http://www.Diamonds.net. March 9, 1999.

——. 2001. "Nicky's Big Deal—Rapaport Analysis." http://www.Diamonds.net. March 2, 2001.

"Rapaport: Trade Alert: 03/19/99." http://www.Diamonds.net. March 19, 1999.

Ratliff, J. D. 1959. "Antwerp's Glitter Street: World Diamond Center." *The Reader's Digest* 74: 225–30.

Rimer, Sara. 1999. "Older People Want to Work in Retirement, Survey Finds." *New York Times*. September 2, 1999: A10.

Roberts, Brian. 1972. *The Diamond Magnates*. London: Hamish Hamilton.

Rosen, Norma. 1992. *Accidents of Influence: Writing As a Woman and a Jew in America*. Albany: State University of New York Press.

Ruby, Walter. 1996. "Eli Izhakoff: Polishing the Image." *Lifestyles* 26, no. 142: 38–40.

Sacks, Karen. 1979. *Sisters and Wives: The Past and Future of Sexual Equality*. Westport, Conn.: Greenwood.

Sacks, Maurie, ed. 1995. Introduction. *Active Voices: Women in Jewish Culture*. Chicago: University of Illinois Press.

Sanjek, Roger, ed. 1990. *Fieldnotes: The Makings of Anthropology*. Ithaca: Cornell University Press.

——. 1998. *The Future of Us All: Race and Neighborhood Politics in New York City*. Ithaca: Cornell University Press.

Savishinsky, Joel S. 2000. *Breaking the Watch: The Meanings of Retirement in America*. Ithaca: Cornell University Press.

Scarisbrick, Diana. 1998. "The Diamond Love and Marriage Ring." In *The Nature of Diamonds*, edited by George E. Harlow. Cambridge (UK): Cambridge University Press (in conjunction with the American Museum of Natural History).

Schmeisser, Peter. 1989. "Going for the Gold: Harry Oppenheimer's Empire." *New York Times Magazine*. March 19, 1989: 32–42.

Schumach, Murray. 1981. *The Diamond People*. New York: W.W. Norton.

Sered, Susan Starr. 1994. "'She Perceives Her Work to be Rewarding': Jewish Women in a Cross-Cultural Perspective." In *Feminist Perspectives on Jewish Studies*, edited by Lynn Davidman and Shelly Tenenbaum. New Haven: Yale University Press.

Shigley, James E. and Thomas Moses. 1998. "Diamonds as Gemstones." In *The Nature of Diamonds*, edited by George E. Harlow. Cambridge (UK): Cambridge University Press (in conjunction with the American Museum of Natural History).

Shor, Russell. 1993. *Connections: A Profile of Diamond People and Their History.* Ramat Gan: World Federation of Diamond Bourses.

———. 1996a. "Angola's Floodgates Reopen." *New York Diamonds* 35: 42–44.

———. 1996b. "DDC Market Week 'Unqualified Success.'" *New York Diamonds* 37 (October/November): 60–61.

———. 1996c. "GIA Colored Diamond Certificates Draw Mixed Reviews." *New York Diamonds* 35: 70–72.

———. 1996d. "Rapaport: 'Don't Blame Me for Low Diamond Profits.'" *New York Diamonds* 37 (October/November): 34–38.

———. 1996e. "Russia Stays, But Argyle May Go: Special Report." *New York Diamonds* 35: 24–40.

———. 1997a. "Industry Report for CSO Stresses New York's Strength." *Diamond World Review* 96: 64–66.

———. 1997b. "New York Manufacturers Seek a Wider Role." *New York Diamonds* 39 (March/April): 36–38.

———. 1997c. "Reinventing 47th Street." *Diamond World Review* 96 (March–April): 70–73.

———. 1998. "Two-Tier Wars May Be in Offing." *New York Diamonds* 44 (January): 44–48.

Silverman, Myrna. 1988. "Family, Kinship, and Ethnicity: Strategies for Upward Mobility." In *Persistence and Flexibility: Anthropological Perspectives on the American Jewish Experience*, edited by Walter P. Zenner. Albany: State University of New York Press.

Spar, Deborah L. 1994. *The Cooperative Edge: The Internal Politics of International Cartels.* Ithaca: Cornell University Press.

Spiegel, Maura. 1998. "Hollywood Loves Diamonds." In *The Nature of Diamonds*, edited by George E. Harlow. Cambridge (UK): Cambridge University Press (in conjunction with the American Museum of Natural History).

Starr, Roger. 1984. "The Real Treasure of 47th Street." Editorial in *New York Times.* March 26, 1984.

Sugden, Robert. 1986. *The Economics of Rights, Cooperation and Welfare.* Oxford: Basil Blackwell.

Teitelbaum, Richard S. 1991. "Hard Times for Diamonds." *Fortune* 123, no. 8 (April 22): 167–78. "The Rocky Road to Survival." 1994. *The Economist 331*, no. 7860:45.

"The Truth about Profits." 1998. http://www.Diamonds.net. December 7, 1998.

Turrell, Robert. 1987. *Capital and Labour on the Kimberley Diamond Fields, 1871–1890.* New York: Cambridge University Press.

Twersky, Boruch and John Myerhoff. 1993. "Antwerp Certs vs. GIA's." Letters to the Editor. *Rapaport Diamond Report.* March 5, 1993: 2.

"UNITA Smugglers Outdo Angolan Government." 1999. http://www.Diamonds.net. August 23, 1999.

Wagner, Thomas. 1999. "Asian Markets Boom Two Years After Facing Grave Financial Crisis." *Providence Journal.* July 3, 1999: B3.

Weber, Lauren. 2001. "The Diamond Game, Shedding Its Mystery." *New York Times.* April 8, 2001: section 3, p. 1.

Weinberg, Sydney Stahl. 1988. *The World of Our Mothers: The Lives of Jewish Immigrant Women*. Chapel Hill: University of North Carolina Press.

Weisser, Michael R. 1989. *A Brotherhood of Memory: Jewish Landsmanshaftn in the New World*. Ithaca: Cornell University Press.

Westwood, Sallie. 2000. "'A Real Romance': Gender, Ethnicity, Trust, and Risk in the Indian Diamond Trade." *Ethnic and Racial Studies* 23: 857–70.

Whitley, Edward. 1992. "Falling Like a Stone." *The Spectator* 269, no. 8567 (September 19): 9–11.

Worger, William H. 1987. *South Africa's City of Diamonds: Mine Workers and Monopoly Capitalism in Kimberley, 1867–1895*. New Haven: Yale University Press.

Yogev, Gedalia. 1978. *Diamonds and Coral: Anglo-Dutch Jews and Eighteenth-Century Trade*. New York: Holmes & Meier Publishers.

Zenner, Walter P., ed. 1988. *Persistence and Flexibility: Anthropological Perspectives on the American Jewish Experience*. Albany: State University of New York Press.

Zenner, Walter P. 1991. *Minorities in the Middle: A Cross-Cultural Analysis*. Albany: State University of New York Press.

——. 1998. "The Ethnography of Diaspora: Studying Syrian Jewry." *Contemporary Jewry* 19: 151–74.

Index